The Birth of Opal

RICHARD SHEPHERD

DEDICATION

This Book is dedicated to my four lovely daughters:

Jessica

Kimberley

Rebecca

Kayla

"For I know the plans I have
for you." Declares the Lord,
"Plans to prosper you and
not to harm you, plans to
give you hope and a future."
[Jeremiah 29:11]

CONTENTS

LIST OF TABLES

LIST OF FIGURES

ACKNOWLEDGMENTS

A special acknowledgement goes to Byron Deveson of Canberra, whose work helped me to understand the formation of sedimentary opal in the Australian environment in a much deeper way.

His work on mound springs and the way they connect with opal formation in Australia was truly a unique perspective and one which deserves honor and I hope I have given him the credit he deserves.

.

1 INTRODUCTION TO OPAL

1.0 General Information

Since the earliest of times, mankind has been fascinated by the variety colour and beauty of opal. Some opal contains tiny flecks of colour, others exhibit bright flashes of colour, some with pin fire whilst the background colour could be clear as in fire opal, or black or white, with the most common being blue. It is the variety of colour, the combinations, the flashes and the changes in colour that make opal one of the most beautiful and sough after gems of all times.

1.1 A Short Opal History

Opal has been known since ancient times, and has been considered a gem stone for over 4,500 years, having been admired by the Egyptians and highly esteemed by the Romans and other civilizations including, the Aztecs, the Egyptians, the Indians and the Arabs.

The following is about two ancient writers and their thoughts about opal:

- Onomacritus an ancient Greek poet wrote
  "the delicate colours and tenderness of the opal reminded him of a loving and beautiful child."

- Pliny the elder wrote
  "for in them you shall see the fire of the Ruby, the glorious purple of the Amethyst, the sea green of the emerald, all glittering together in an incredible mixture of light."

1.2 Origins of the name

Many of the early civilizations had their own words for opal. In the ancient Sanskrit the word for opal was 'upala' with the simple meaning of 'precious stone'. The early Greek had a few words for opal, the first word was the same word used for eye 'paederes' and carrying the secondary meanings of 'child' or 'favorite'. The second word was 'opalus' or 'Opallios' meaning to see a change of colour. The Roman word was 'opthalmus'. The Romans believed that opal was symbolic for purity and hope.

1.3 Opal as a Gemstone

Opal has been revered as a gemstone from the earliest of times. According to Pliny the elder, opal was above all gemstones, the most highly prized in the entire empire.

Mark Anthony lust for opals was legendary.

Opals have featured in the treasures of many royal collections. Queen Victoria was said to give them as wedding presents. The Holy Roman Emperor featured an opal in his crown.

In the Middle Ages it was thought to be a lucky charm, yet in modern times it is considered unlucky for those who are superstitious.

1.4 Opal Origins

Opal has been mined in many different countries around the globe including:

- Mexico
- Nevada USA
- Brazil
- Honduras
- Nicaragua
- Guatemala
- Japan
- Ireland

Recently opal deposits have been discovered in Ethiopia, some of the opal is reported to be brittle, whilst some is of good quality.

However by far the largest producer of opal in the world is Australia, which produces between 90-95% of the world's supply of opal. This opal is mostly mined within the Great Artesian Basin area, thought once to have been an inland sea. Australian opal differs to most other opal as it is of sedimentary origin not volcanic like most other deposits.

1.5 Opal in Australia

It was a German geologist by the name of Johannes Menge who first discovered opal in Australia, near Angaston SA in 1849. The 1880's saw the discovery of the Lightning Ridge fields and Bolder opal in Queensland. Many other discoveries soon followed including White Cliffs in NSW, Opalton in QLD, and Coober Pedy, Andamooka and Mintabie in South Australia.

Today the finest opal is extracted from South Australia [Coober Pedy, and Andamooka], Queensland, and New South Wales at Lightning Ridge. Other countries where opal is mined, Japan for white opal, Mexico & Honduras produce fire opal, whilst India, New Zealand and the west of the United States produce a variety of opal types.

1.6 How Does Opal Form?
This publication is an investigation into the natural formation of Opal. In this search we will need to uncover as many clues as possible to establish a cause for opal formation. The search will take us through looking for clues in the chemical makeup of the opal, its environment, and the clays in which it is formed and even into the method of making synthetic opal. The author already has specific views as to how opal forms and seeks to find the evidence to prove them.

A good starting point would be to give the traditional definition of opal and then to give my arguments as to why it is wrong and how this reflects on the natural opal formation process.

1.7 Traditional Definition of Opal

Opal is said to be an amorphous [without specific shape] non-crystalline gem mineral which has solidified from Silica gel, which is a type of liquid silica which is deposited in cracks and cavities left by decaying vegetable and animal matter. The chemical composition is $SiO_2 + H_2O$. High grade opal contains between 6-10% water.

1.8 Modern Definition of Opal

The modifications that I would add to this definition are that it is very definitely crystalline in nature, this is proven out by electron microscope pictures called micrographs. The silica spheres from which it is assembled are a type of micro crystal that range in size from 150-400nm.

Opal is a Polymer; produced by chemical reaction, that is; it is a result of a polymerization process. It is not just silica gel deposited in cracks. The polymerization process is a growth process which produces the veins or seams of opal which can be horizontal or vertical. Opal growth in veins is evidence of a polymerization process. The opal grew when the surrounding clays were soft and wet.

The opal spheres begin as minute colloidal particles with sphere sizes between 1-20nms in size and grow by a process called Ostwald ripening to within the size range of 150-400nm [1500-3500 Angstroms]. Science has largely ignored these important chemical and scientific processes due to its lack of knowledge of Silica and its low temperature chemical reactions [Science is rapidly beginning to catch-up in this area.].

During the process of Polymerization as the particles reach their ultimate size settling takes place whereby the particles arrange themselves in a close packed ordered array of silica spheres [See CH: 4 Polymerization] .

The Australian Opal Industry supplies 95% of the world supply of Opal. In 1996-97 the Australian industry sold around $120 million worth of opal.

However really valuable opals are quite rare, representing only about 5% of the total opal production. That leaves 95% which are of lesser value, many of which are almost worthless being mostly potch or common opal rather than valuable precious opal. It is for this reason, scarcity of premium quality precious opal that it commands such high prices. Good quality black opal can fetch prices as much as $15,000 per carat. This is higher than the price of Gold per carat, and even more than most Diamond.

1.9 Pricing Opal

Recently there has been a change in the way that opal is being categorized. The Australian Opal & Gem Industry Assoc. [AOGIA] has introduced a system of categorizing opal under 3 main headings based upon the opals body tone. These 3 three headings are:

- Black Opal N1, N2, N3, N4
- Dark Opal N5 N6
- Light Opal N7 N8 N9

Black opal is the ultimate in opal value especially if it contains some orange or red. Black opal is thought to contain some black potch covered by a veneer of precious opal. Others believe that the black colouring comes from inclusions of organic carbon or manganese dioxide. Reds and yellows within black opal may come from impurities of iron oxides which are natural colouring agents. Most black opal comes from Lightning Ridge in NSW.

In similar fashion Dark Opal is thought to have a dark body tone arising from a dark potch background. Dark opal has a play-of-colour over this dark body tone. Dark opal is next in line in value to Black opal.

Light Opal can have a body tone anywhere between colourless to medium greys. Opals with very milky body colour are the only ones that should be called white opal. Most of the white opal in Australia comes from Coober Pedy in South Australia.

1.10 Colour value

Blues in black opal are not only desirable but are also attractive ranging in hues from light blue to the deep sapphire and electric blues. Blues are the most common and therefore fetch a lower price.

Green on black is not as common as and therefore more valuable than the blues. If the green also contains some orange or red then the value will increase even more.

Yellow is quite rare in black opals but sometimes can be seen as a clear lemon yellow. More often it may appear as a golden colour or may be combined with green or perhaps orange.

Orange is the second most desirable colour in black opal. It will really increase the opal value if it is a deep rich orange which saturates the opal. If combined with small amounts of green or red it becomes a very valuable gem stone.

Red is the rarest and most highly prized of all the colours in black opal. Reds can vary from wine red to crimson. Red gives opal its highest value and is enhanced even more if combines with, green orange and blue. This is the rarest and finest of all the opal gems.

Factors which affect the value and pricing of opal are:

- Body Tone
- Brilliance
- Pattern
- Colour Bar thickness
- Faults
- Play-of-colour
- Cutting and Polishing

Body Tone has already been mentioned. Brilliance or the brightness of the colours affect price. The brighter the colours the more valuable the opal. Different colour patterns can vary considerably and also affect value of the opal. Colour bar thickness is a measure of the thickness of the precious opal, which is often just a thin veneer over potch.

Raw or uncut opal is worth a fraction of the price that it could command if properly cut and polished and free from faults such as cracking or crazing. Crazing are micro fine cracks which render the opal almost worthless as do cracks to the face of the opal.

Sometimes opal also contains inclusions which can also have a negative impact on value. Some types of inclusions are sand, gypsum or other unwanted minerals.

Most small opal that is cut and polished is used in the jewelers' trade in pieces like rings, ear rings, bracelets, and pendants. Larger pieces often find use in brooches, clasps or other peices of jewellery.

Accepted Theories of Opal Formation

1.11 Current Theories

There are now in fact three main accepted theories of opal formation, these are:

- The Microbial Model
- The Weathering Model
- The Syntectonic Model

It may be surprising but most of these models are only a few years old, as the models that existed before these have been discarded as unsupportable.

1.12 The Microbial Model

Opal host rock, often a type of clay stone, such as Bulldog shale, and even ironstone contain high levels of fine particles of fossilized organic matter. Various types of fossil microbes have been identified within opal, mostly bacteria of the aerobic type, that is the oxygen breathing variety. These tiny microbe inclusions in the opal structure are not visible to the naked eye but seen through a microscope they appear in abundance. This model puts forward the concept that the feeding and waste cycles of the living microbes provide suitable conditions both chemically and physically required for opal creation. In essence the claim of this model is that the microbes are responsible for the building of opal.

1.13 The Weathering Model [Ref: 15]

This model is also referred to as "The Deep Weathering Model".

"During the Tertiary period the rocks which now contain opal were subject to significant weathering. Over time small amounts of silica tended to be leached from sandstone layers by water which then passed through the rock until it became trapped by underlying layers of relatively impermeable clay stone. Under the right chemical and physical conditions opal was precipitated from the water in porous areas or voids.

There is a tendency for opal to be found close to faults in the rock layers and near " blows" (disturbed ground). If this model is correct then these faults and blows would have to provide paths for water containing silica to flow along." [Matthew Goodwin, Ref: 15, 2002].

1.14 The Syntectonic Model

This model suggests that heated pressurized water from underground deposited opal. This water preferentially followed faults and blows, depositing the opal nearby.

Some scientists even went as far as to try to pin opal formation to the theory of evolution, suggesting time spans for formation and age of opal deposits in the millions of years. However this type of speculation was not only self serving but under recent scrutiny has become redundant in light of the following facts.

- Opal can be formed in the Lab. In weeks or months not years, certainly not millions of years. It is the result of chemical reactions and the right conditions.
- Opal breaks down over time [this is linked to silicic phases and the silicon cycle], it is unlikely to exist in opal form much past 10,000 years.
- Byron Deveson carried out carbon 14 testing on lightning ridge opal, which dated it at approx 4000 years.

A close examination of each of the above models will reveal that each has strengths and weaknesses. It may be found in time that a mixture of all three models maybe required to ultimately explain opal formation in nature.

1.15 A New Model Arises [Mound Springs]

Byron Deveson has recently proposed a new model for opal formation.

This new method is based upon the syntectonic model but also accommodates roles for bacteria and weathering fluids. This new model proposes Mound springs as the environment in which sedimentary opal forms.

> *"A new model for the formation of precious and potch opal is proposed. The essential components of this model are mound spring waters of appropriate chemistry; a mechanism whereby the physio-chemical properties of this water are changed so that suitable silica spheres, and then linear chains of these spheres, are formed; and suitable voids that are lined with clay that can act as a semi-permeable membranes to concentrate and purify the silica sol by ultra filtration and dialysis."*[Byron Deveson, Ref: 14, page 1]

Some of the mound springs are now extinct but many remain and are features of the GAB. These mound springs are extensive. Active and extinct mound springs are common in many sedimentary opal producing areas in South Australia, NSW and Queensland: including Lightning Ridge, Mehi, Lila springs, Nymagee, White Cliffs, Andamooka, Coober Pedy, and the Eula/Yowah districts.

There is strong evidence to suggest the involvement of mound springs in the formation of sedimentary opal in Australia.

The impurities [inclusions] in opal are consistent with those found in carbonatites or natrocarbonatites from mound spring waters. These springs are often surrounded by broken silcrete boulders which are known to form as capping on opal. It is thought that silcrete forms simultaneously with opal.

1.16 The Redundant Links to Evolution

Geologists and scientists would have us believe that opal was formed by dissolved silica being transported by acidic water flowing through the sediment and getting caught in cracks and crevices where it remains and settles forming regular shape and sphere sizes. This colloidal silica thickens as water carries more dissolved silica into these same cracks. Additions of silica leads to gel formation eventually drying out to form opal. Some sources even state that this process takes millions of years.

This scenario presents us with numerous problems:

- How could silica gel remain in cavities unaltered for millions of years. Extremely unlikely, if you were a betting man you would call this a long shot.
- Silica gel itself will not form precious opal, a chemical process is involved.
- Does not account for bolder opal
- Does not account for volcanic opal
- Does not take into account polymerization, which explains the chemically bound water in the opal structure
- Time span credited with such a process is totally unrealistic, we already know that opal can be made in the laboratory in under a year.

Chemical reactions occur quickly not over millions of years. Millions of years as a timeframe for single chemical reactions are unknown to modern science except for the Fusion reaction.

- Pseudomorphism not explained properly by such theory, the terminology used is very general. Replacement and impregnation are outdated explanations which don't do justice to the process. Ion exchange and polymerization are more likely the mechanism.
- Does not account for horizontal opal veins. Horizontal veins are caused by a growth process, and the only chemical growth process that I know about for inorganic matter is polymerization.
- Physical evidence has been ignored, the fact that the silcrete forms as a byproduct of the opal polymerization process, indicates oxide catalysts have been involved, but have not been investigated.

Debunking Opal Myths

1.17 Myth Busting

1. Opal is formed by deposition of dissolved silica. This would seem to imply polymerization by addition, building up layer by layer. However this does not work. You can do the same by using Sodium Silicate and making Silicic acid and adding more Silicic acid but opal does not result. Not all silica gels form opal.

2. Opal was formed millions of years ago, implying that all the opal was formed in the distant past. Proven wrong by:
   - Len Cram [See: The World of Opal]
   - Russian Opal Synthesis
   - Recent Opal formations: Fence Posts [5 years]
   - Opal Miners cat

   The millions of years idea has no evidence to support it, very unlikely and does not take into account earth movements, quakes and minor quakes would unsettle this process, such a long period allows for contamination to occur on large scale.

3. Opal is not crystalline. You may have heard it said that opal is not crystalline because it is amorphous, meaning without shape. The fact however is that opal is composed of micro crystals. This has been proven by CSIRO Scientists using the electron microscope. Recently these crystals have become known as photonic crystals because of their ability to bend light.

The changing nature of opal itself testifies against evolutionary theories. Opal changes over time due to loss of water and crystal morphology. Over long periods of time the crystal structure of opal changes with opal degrading to chert. It is possible for opal to remain as opal for thousands of years but its tendency to change casts a long shadow over notions that it could exist in the same state for millions of years. The chance that opal that exists today existed as opal even one million years ago is so far remote that it lies in the realms of the impossible.

Related Topic

Geography; Geology; groundwater; bore water; aquifers; hydrology; water resources; inland sea; opal; Australian opal towns; oil deposits; plutonic water, sedimentology. Opal Myths; Petrification; Opalization; Polymorphism; Photonic crystals;

Silicification; Acid-sulfates; Sulfuric acid alteration; photolysis; photonic crystals; current opal theory; geology of opal; photo catalysis; photochemical reactions; opal environment; hydrothermal alteration; Sedimentary opal.

Fossicking; Opal; Mining; Licenses; Coober Pedy; Lightning Ridge; Opal types; Opal environment; Precious opal; potch (common opal).

Opal towns;

Related Websites

Geology of the far North of South Australia

Library.adelaide.edu.au/guide/sci/Geology/farnorth.html

The Great Artesian Basin (report by Peter Lewis)

www.abc.net./landline/I1101099.htm

Online Opinion by Lance Endersbee - source of water

Pandora.nla.gov.au/pan/10635/20010706

www.onlineopinion.com.au/2001/may01/endersbee.htm

Dinosaurs and the Great Artesian Basin

www.qmuseum.qld.gov.au/features/dinosaurs/elliot/2002/par
ticipants.asp

Formation of opal

Source: http://www.opal-shop.com/mineforopal.htm

Further Reading

Investigations of the Geology and Hydrology of the Great
Artesian Basin, 1878-1980

www.bossintl.com/bookstore/index/book/2010.html

The World of Opal
Author: Allen W. Eckert
Publisher: Wiley 1997.

Bibliography: Chapter 1

1. *Opal value*

Opals Down Under

http://www.opalsdownunder.com.au/articles/fields.htm

2. *Opal Cutting*
Clifford Coan
American Opal Society
http://www.opalsociety.org/opal_cutting.htm

3. *Opal Types*
Colonial Opal
http://www.colonialopal.com.au/types.html

4. *The Australian Opal Industry*
http://www.costellos.com.au/opals/industry.html
http://www.costellos.com.au/opals/valuing.html

5. *Parched Earth Opals*
Opal Types
Sandra McCondra

6. *Opalfields Information*
http://www.opalfields.com.au/information.htm

7. *Opal Colour*
The Opal Hut
http://www.opalhut.com.au/opal.htm

8. *Encyclopedia Britannica*

Vol 8, Pg 960

Published: Chicago USA, 2002.

9. *Opal: South Australia's Gemstone*

Editors: L.C. Barnes; T.J. Townsend; R.S. Robertson; and D.C. Scott

Published by:

Dept Mines & Energy Geological Survey of South Australia 1992

Library Ref: 622.

10. *Opals of the Never Never*

By: Robert Haill

Published by: Castle Books, Australia 1981

11. *Black Opal fossils of Lightning Ridge*; by Elizabeth Smith; Published by Kangaroo Press 1999).

12. *A Fossickers Guide to Gemstones in Australia*

By: Nance & Ron Perry

Published by: Reed Books Australia 1997

13 *Columbia Encyclopedia*

New Illustrated Edition
Vol 10. Pg 2818 [Great Artesian Basin]

Columbia University Press, NY 1978.

14. *A new Opal Model*

Mound Springs as Opal Source

Byron Deveson

Canberra 2006

15. *Geology of the Opal fields in the Lightning Ridge region*

Subject: How Opal was formed

www.wj.com.au/mining/Lrgeology.html

Matthew Goodwin [2002]

16. *Opal Genesis*

Lightning Ridge Miners Assoc.

LRMA Newsletter 15 May 2002

http://www.wj.com.au/newslrma/lrma/lrma12.html

17. National Opal

www.nationalopal.com/about_opals/history.html

Subject: Opal History

Date searched: 28/3/2012

18. Graham Black Opal

www.grahamblackopal.com

Subject: Opal History

Date Searched: 28/3/2012

19. Opal Minded

www.opalminded.com

Subject: Opal History

Date Searched: 28/3/2012

THE BIRTH OF OPAL

2 SEDIMENTARY OPAL ENVIRONMENT

The Great Artesian Basin

2.0 The Opal Environment

The first clue will be discovered from the natural opal environment. Knowing where opal is formed, helps us to understand the natural processes involved. As the formation of Opal is tied to its immediate environment. The processes that formed the opal environment must also have impacted the formation of opal. Let us then look at the landscape in which opal has formed, The Great Artesian Basin.

The Great Artesian Basin was thought to have once been a great inland sea which is evidenced by the presence of sea shells found throughout this vast area.

"Most of the sediments of the Great Artesian Basin are marine sediments. Thus, the uppermost strata in the sequence must have been below sea level at the end of deposition. This indicates that the entire thickness of the sequence of saturated and partially consolidated sediments would have been below sea level when the uppermost sediments were being deposited. It follows that the base of the sedimentary series must have been at a level of 4000 metres or more below sea level, say the same level as the present level of the ocean floor of the Tasman Plain.

But the entire sequence is now elevated and consolidated. That indicates that there have been substantial changes in the elevation of much of Australia since the beginning of deposition of sediments in the Great Artesian Basin.

The strata of the Great Artesian are relatively undisturbed by faulting, and are essentially flat for hundreds of kilometres, north and south, and east and west. It follows that all that vast area of Australia, which was underwater when the Basin sediments were being deposited, has since moved up and down as one huge block, and over a vertical range of at least 4000 metres. There was probably a succession of movements, dating from the Triassic or earlier, and successive periods of sedimentation, uplift, erosion and subsidence."

(Quote Source: Ref 63; L.A. Endersbee.)

The most famous of the Australian opal mines exists within the vast expanse known as the Great Artesian Basin. This area is so rich in opal that it accounts for something like 90% of the world production of opal.

The Great Artesian Basin extends from the Gulf of Carpentaria, in far northern Queensland through central western NSW and well into South Australia. It contains the opal towns of Coober Pedy, Mintabie, and Andamooka in South Australia, Lightning ridge in NSW and Quilpie and

Opalton in QLD. The expectation that great undiscovered opal riches are yet to be uncovered in this huge expanse seems very likely.

If you have a good look at a map of the Great Artesian Basin you will begin to understand not only the enormous size of this area but also the potential that more great discoveries are beckoning. This whole area was formed under the same conditions and this is the great indicator that more strikes maybe awaiting some intrepid explorer.

2.1 Australian Opal Towns

Some of the most famous of the Australian Opal towns are Coober Pedy, Lightning Ridge and Andamooka. Coober Pedy is the largest producer of Opal in Australia. Lightning Ridge and recently Andamooka are famous around the world for the beautiful black opal that is recovered from these regions.

The large majority of the well known opal towns exist within the Great Artesian Basin. The significance of this fact is huge because it acts as a guide to us about the conditions under which Australian opal has formed. It would not be surprising to find that opals in other parts of the world were formed in very similar conditions and terrain.

In order to gain some perspective on the landscape and the conditions in which opal formed it is necessary to view a map of the Australia featuring The Great Artesian Basin.

Map shows expanse of the Great Artesian Basin

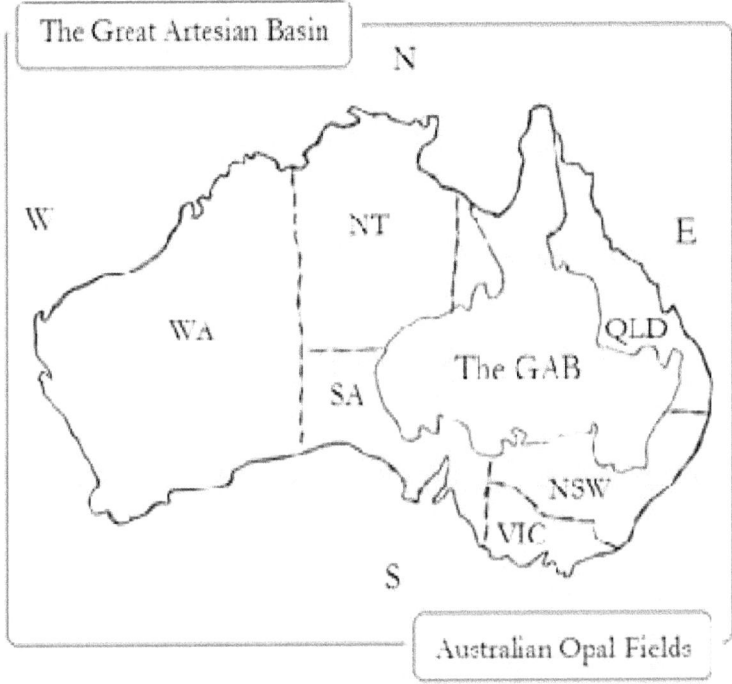

The area of the Great Artesian Basin is clearly labeled, it covers an area of approx 20% of Australia, 670,000 sq miles or 1,735,300 sq kms.

Figure 2. Australian Opal Towns

Australias Opals Towns

Opal Towns:

Queensland:
1. Kynuna 2. Opalton 3. Jundah 4 Kyabra 5. Duck Creek
6. Yowah 7. Koroit.

New South Wales:
8. New Angledool 9. Lightning Ridge 10. Grawin 11. White Cliffs

South Australia
12. Myall Creek 13. Andamooka 14. Coober Pedy 15. Mintabie
16. Granite Downs

Western Australia
17. Kurnalpi

It is interesting to note that only two of Australia's opal towns fall outside the GAB.

The conditions which exist in the Great Artesian Basin are an arid landscape with summer temperatures between 35 - 50C for most of the summer. This whole area is a desert type environment with a low annual rainfall. It is hot and humid for most of the year.

Opal Indicators or Markers

TABLE 1 – Opal Indicators

Iron oxides	Are also products of weathering and find their way into sandstones and evaporates, as wind and water transport them. Iron oxides / hydroxides (result of FeS_2 reactions), are associated with most of the different types of opal
Bulldog shale	Is another product associated with opal. The shale often underpins the opal. Opal can form in mudstones siltstones, sandstones and even limestones. Mostly in the former than in the latter. However in Australia it forms mostly in Bulldog shale which many people mistake for sandstone.
Alunite	(Sulfuric acid product of Kaolinite), found in such abundance at Coober Pedy that it is mined there.
Glauberite	(Sulfuric Acid Product); abundant in the opal fields within Australia.

Silcrete	All sedimentary opal formations are or have been capped with silcrete [CH 5].
	Opal formations which lack silcrete are due to weathering, earth movement, in some cases by man. Silcrete is an extremely good sign.
Glug	Peachy orange potch often Yowah nuts, sometimes also contain precious opal.
Celadonite	Blue Clay, Result of Acid Sulfate reactions.

2.2 Where to look for Opal

Opal indicators, sometimes known as markers or tracers are those mineral formations that are associated with Opal. Opals sometimes form in ironstone, which is a sandstone impregnated with Iron. Bolder opal of Queensland is formed in Ironstone. Most Australian opal is found in association with iron. You will realize this when you read through descriptions of different types of opal found in Australia and its environment.

Soils of the GAB

2.3 Opal host soils

The main soils of the Great Artesian Basin are Sandstones, Clays, Bulldog Shale, Carbonatites, Ironstone, Silcrete and Evaporites [salts].

The major processes by which minerals are formed, decomposed and altered are called weathering.

Weathering provides the starting materials for the opal formation process. Opal is formed from very fine grained dust particles. Theses are the particles that are contained in clays and feldspars.

The actual process of weathering is not a single process but many different processes classified under the single heading of weathering.

There are two main types of weathering which are:

- Physical Weathering
- Chemical Weathering

These two types of weathering break down existing rocks, often particle by particle, releasing elements and minerals for further interaction, decomposition, or dissolution. The weathering process can result in the deposition of large quantities of a mineral in almost pure form, or it can result in the formation of new rock structures.

2.4 Minerals & Rocks of GAB

Common mineral strata of the GAB include:

- Sulfides; predominantly FeS2 Iron Sulfide
- Carbonates, mostly CaCO3 Calcium Carbonate
- Sandstones & silicates (Olivine very common)
- Clays; mostly Kaolinite & Bentonite
- Evaporites; salts; Chlorides; sulfates
- Shale
- All soils contain Organic Matter

Decomposition Chemicals:

- Sulfuric Acid & HCL
- Carbonic Acid
- Fenton Reaction
- FeS_2
- Alkali
- Sulfates
- Phenols

Chlorides (salts) help to disperse the clays to allow chemicals to react with the whole clay volume.

2.5 Common Rocks & Minerals

Quartz and Sandstones

The origin of most of the silts and sand is from the breakdown of quartz.

Found in large quantities in sandstone and granite, these are often the source of quartz decomposition. It is a very resistant material which when decomposed form the principle soil ingredients form the sand and silt type soils. Much of the Great Artesian Basin is Sandstone.

Ironstone

A type of sandstone that has large amounts of iron scattered throughout the structure of the sandstone.

Bulldog Shale

This is a true shale built up from silts but does resemble sandstone and is often mistaken for sandstone. Shale often contain organic materials, some even contain oil. It is highly significant as a major host rock to opal in Australia.

Haematite

This is a red iron (ferric) oxide that is often used as a colouring agent for cements. Due to the fact that it is very resistant to change, it results from extreme weathering. You may notice it as a red or pink colouring in rocks and soils. It is known to form in duricrusts, which are natural cements, which form in arid climates. Duricrusts are related to silcrete, or hardpan soils.

Mica

Flaky minerals originating in many igneous rocks. Some of these are resistant to weathering, such as the muscovites. Others broken down, these include biotite.

Biotite [Mica] is attacked by acid releasing soluble silica which undergoes continuous reprecipitation to form colloidal particles, which aggregate into spherical particles. These could be transported as a sol to the place of deposition. It has been noted that abundant sulfates are found in the environment of opal deposition (3% soluble sulfates and chlorides → both salts) [Ref 6].

Clay minerals

Clays are for the most part derived from the decomposition of the feldspar minerals. Particles are often flaky resulting in them having large surface areas. These decomposed feldspars are the main particles within the clay structure. Some clay also contains silt sized particles.

> *"Clay minerals are produced mainly from the chemical weathering and decomposition of feldspars,*
>
> *Such as orthoclase and plagioclase, and some micas. They are small in size and very flaky (platy) in shape"* (Ref 1; Link to Rocks and Minerals).

Other sources state silica particles in clay are spherical.

Carbonatites and Natrocarbonatites

These igneous rocks, originating from volcanic lavas, contain more than 70% by volume of carbonate minerals, the major elements being; CaO, CO_2 or Na_2O with inclusions of silicate minerals associated with the following rocks:

- Pyroxene
- Olivine

- Nepheline
- Feldspathoids

May also contain rarer elements such as Ba, Cs, Rb and smaller quantities of Hf, Zr, and Ti. They may also contain the rare earth elements: copper, iron, phosphorus, niobium, uranium, thorium, barium, fluorine, and zirconium.

Some of these elements have been identified as nucleating agents for opal genesis.

Natrocarbonatites are very similar to Carbonatites but are a rich source of the alkaline minerals Sodium and Potassium.

Carbonatites are often found in mound springs walls, with traces in the mound spring waters. The alkaline nature of these rocks is believed to be the main source of alkalinity in these waters, and responsible for the impurities found in opal. According to Byron Deveson they are a very close match.

Pyroxene

Pyroxene is a group of rock forming silicate minerals which contain calcium, iron and magnesium in varying degrees, often found in igneous and metamorphic rocks. They have a general formula $XY(Si,Al)_2O_6$ where X may represent Calcium, Sodium, Iron, or Magnesium and Y represents ions of smaller size, such as Chromium, aluminium, Iron+3 or manganese.

Jadeite $(NaAlSi_2O_6)$ the more valued form of jade, is a pyroxene as is Augite the most abundant of the group.

An interesting piece of trivia is that Pyroxene and Olivine comprise the bulk of the upper mantle of the earth. This is also an indication of the abundance of these two groups of rock forming silicate minerals.

Olivine Species [Ref:43]

Olivine is a very common mineral not just in Australia but worldwide. It is also one of the most abundant minerals in the earth's mantle. In the Australian opal fields Olivine is also a common mineral. Olivine is often associated with artesian basins. Olivine is of volcanic origins being one of the first minerals to crystallize out of the magma, and it is possible that it could be present in volcanic ash. Due to weathering olivine can also be found in abundance in the sedimentary environment.

Olivine is an orthosilicate or orthorhombic Nesosilicate having an xSiO4 structure enabling them to dissolves completely [congruent dissolution] in HCL to give almost 100% monosilicic acid.

List of Olivine species:

- <u>Fayalite</u> (*Iron Silicate*)
- <u>Forsterite</u> (*Magnesium Silicate*)
- **Glaucochroite** (*Calcium Manganese Silicate*) *
- **Kirschsteinite** (*Calcium Iron Silicate*) *
- **Laihunite** (*Iron Silicate*)
- **Liebenbergite** (*Nickel Magnesium Silicate*)
- **Monticellite** (*Calcium Magnesium Silicate*) *
- <u>Olivine</u> (*Magnesium Iron Silicate*) $MgFeSiO_4$
- <u>Tephroite</u> (*Manganese Silicate*)

Nepheline

Nepheline belongs to a group called the Feldspathoids. It is yet another silicate, a Sodium Potassium Aluminium Silicate. Its origins are volcanic in nature. It contains Sodium and Potassium in the ratio of 3:1. In nature it may alter to zeolites (Natrolite), Sodalite, Kaolin or compact Muscovite. Nepheline shares a similar trait to Olivine in that it will dissolve completely in HCL with a separation of gelatinous silica (which can easily be stained by colouring matter, such as salts). The True or natural formula is suggested to be $(Na,K)AlSiO_4$ which is an orthosilicate form.

Feldspathoids

This group of minerals is related by their affinity with the feldspar group. These minerals are low in silica but would have formed feldspar if more silica had been present in their originating magmas. They are known as tectosilicate minerals.

Common feldspathoid minerals are:

- Nepheline
- Leucite
- Analcime
- Cancrinite
- Sodalite group

Opal Dirt or Opal Clays

In a clay mineral the elements (oxygen, silicon, potassium, etc.) are spheres arranged in a regular three-dimensional pattern. The spheres are the building blocks of the clay mineral, and the arrangement of the spheres determines the type of mineral. The character of the clay mineral group determines the type of clay and its eventual use. In other words, the clay mineral structure gives us an understanding of its specific properties. [Ref: 5]

Clays then are already composed of the silica spheres necessary for opal production. Opal dirt is essentially clay.

Opal dirt is the dirt from which opal is formed. Normally refers to the following clays:

- Kaolinite (Kaolin) 75%
- Montmorillonite (Smectite rather than Bentonite) 20%
- Illite (Sericite) 5%

[Ref: 4; Elizabeth Smith; pg 48]

Clays are AluminoSilicates, also called Hydrous Oxides. Clays also includes:

- Chlorites
- Feldspars (Alkali AluminoSilicates)
- Vermiculite/ Perlite
- Zeolites
- Oxyhydroxides

Oxyhydroxides are considered by some sources to be clays. Chlorites are a large group with varied formulas, and could act as a valuable source of ions for exchange reactions. Vermiculite and zeolites are extremely good ion exchangers.

Kaolin and Smectite clays are easy to obtain through a ceramics/pottery outlet. Kaolin is the major clay used in ceramics. Bentonite is not pure Montmorillonite clay, despite its impurities maybe considered a substitute for Smectite. Also used in some ceramics applications, however being an expanding clay its major use is probably dam sealing. You may also find these clays though a mineral supplier like UNIMIN. Unimin would also be able to supply illite but the formula requires such a small amount that you may be better off making your own. Bentonite also contains a small amount of illite.

Illite is going to form as an alteration product of the clay and feldspar reactions. It may not therefore be needed as an ingredient.

The other important material needed is K-feldspar which initiates the silicification process. K-Feldspar is a Potassium Aluminium Silicate. K-Feldspar is also available through UNIMIN.

| Soil Reactions of the GAB |

2.6 Soil Resources

There are a number of important soil reactions that occur or have occurred in the GAB. The soils involved in the reactions are chiefly the clays, sulfates [mainly FeS2], carbonatites and evaporates. These reactions are important sources of released Silicic acids and water.

2.7 The Chemical nature of clays

Hydrous oxides are polymeric solids formed by the hydrolysis of certain metals, especially iron, manganese, silicon and aluminium. In aqueous environments they are typically found in amorphous or poorly crystalline forms, as coatings on the surfaces of mineral particles. The importance of hydrous oxides as complexers of metals derives from the presence of hydroxy functional groups at their surfaces, which are able to exchange protons for metals. They may possess high surface areas (of the order of several hundred square metres per gram) and a porous gel-type structure allowing metals to bind to both external and internal surfaces

2.8 Flocculation & Dispersion of Clays

The flocculation process brings together individual clay particles to form floccular aggregates. This is also called coagulation. It is a reversible process but can be relatively permanent depending upon the ions that are present. Calcium and hydrogen ⇑ increase flocculation but potassium and sodium ⇓ cause the reverse, dispersion. Dispersion is the process whereby individual clay particles are kept separate from each other. Clays heavy in sodium content have a dense electric double layer surrounding the ion, causing the clay to remain in suspension. Calcium has the ability to suppress this double layer and initiate flocculation. The effectiveness of flocculation is increased with Tri- and tetravalent ions.

Thus for movement or transport of the clay dispersion is required so that the clay remains suspended and can be easily be transported by moving water.

Dispersants:

- Potassium (K)
- Sodium (Na)
- Other clay dispersants (Tripolyphosphates)
- Gypsum (Calcium Sulfate).

Flocculants:

- Calcium (Ca)
- Hydrogen (H)

2.9 Transportation

Both wind and water can transport Clays. They form sedimentary layers. These are simply referred to as sediments, which imply a deposition of the weathered materials.

Sedimentary layers can be very thick, even up to kilometres in depth and spread over vast distances like the sediments of The

Great Artesian Basin.

Sediments are deposited in water bodies, such as lakes, seas and oceans. The Great Artesian Basin was once a sea and soils in this huge area are almost exclusively sedimentary in origin.

2.10 Illite formation

- Muscovite (fine grains)
- Sericite

The first two clays are not hard to obtain, although Illite can be difficult to get. Illite can be formed in two ways

Two methods that I have discovered are:

- The dissolution of K-Feldspar results in Kaolin taking up the K ions(potassium) causing the precipitation of Illite.
- Kaolin is also known to take up K ions from seawater to form illite. If K-feldspar dissolves in water.
- The altered product of Smectite. Smectite is not difficult to obtain.

Using the first method then, is a very simple way to make Illite.

K-Feldspar dissolved in water (hydrolysis) then adds Kaolin. SIMPLE!!

2.11 Common Clay Reactions in GAB

In reference to the Stober method of opal formation, scientists do not believe that opal could have formed this way. However this author believes that they have missed the point. The Stober method is only a vehicle to SiO_2 (aq); this is a convenience for man. Nature does not require TEOS to form SiO_2 (aq) or colloidal silica. The other main chemicals however would be the same, or very similar. These organic chemicals are derived from the decay of animal and plant matter. Ammonia may be in the form of NH_3 (aq) or in the form of Amines; the solvent could be either Ethanol or Methanol formed from decay.

The nucleation of silica occurs through dissolution of a natural silica source; probably Olivine. Growth of silica particles occurs in the same way; by Ostwald ripening [can be likened to polymerization by addition; seems to be the same mechanism].

In nature the reaction within the clay volume may start as an acidic reaction due to the Acid Sulfates that are present in both the volcanic and sedimentary environments causing large amounts of alumina to be dissolved from the clay. Dissolved aluminium exists only as the Al_3^+ ion which hydrates in the presence of water to form Aluminium hydroxide. The other outcome is that the silica will now be easier to dissolve.

Studies of the formation of volcanic opal have shown that the volcanic fluid that produces opal is Acid Sulfates where they mix with neutral chlorides. This would result in the following reaction:

Sulfuric acid with sodium chloride:

$$NaCl + H_2SO_4 \rightarrow NaHSO_4 + HCl\uparrow$$

$$2\,NaCl + H_2SO_4 \rightarrow 2\,HCl + Na_2SO_4$$

Olivine [Silica Chapter] is very common in the Australian Opal Fields. When Olivine reacts with HCL the result is the congruent dissolution of silica with the other ions forming chlorides. This dissolved colloidal silica could be the precursor to Opal formation. The chlorides may then react as follows:

HCL is a very reactive acid and will combine with free ions to form salts again. When chlorides [NaCl very common] are present with the sulfates and reacts with carbonates, sodium carbonate is formed causing an upward spike in the pH.

2.12 pH Change in clay sediments

Silica dissolution requires acidic pH. Ostwald ripening requires the pH to change to an alkaline pH 7 or above. At pH 8.3 Silica solubility starts to increase and is very high at pH 10 and above.

The H+ ions can combine with any carbonates to form Carbonic acid which is also known to dissolve silica.

> "It has been suggested that the dissolution kinetics of most slightly soluble oxides and silicates is controlled by the concentration of adsorbed charged species at the mineral surface produced by these reactions, in particular by H+ and OH- (e.g., Chou and Wollast, 1984; Knauss and Wolery, 1986; Carroll-Webb and Walther, 1988; Brady and Walther, 1989, 1990; and many others)". [Ref: 8; Wan-Joo Choi].

The H+ and OH- ions are naturally produced in the following clay reactions.

2.13 Common Clay Reactions

> *"The principal cause of alkaline reaction of soils is the hydrolysis of either the exchangeable cations or of such salts as CaCO3, MgCO3, Na3CO3, etc. Hydrolysis of the exchangeable cations takes place according to the following reactions*

Clay micelle (Na x 2) + H_2O $\longleftrightarrow$ Clay micelle (Na + H) Na^+ + OH^-

In this reaction H+ is inactivated by exchange adsorption in place of Na+. The displaced Na does not combine with, or inactivate OH- ions which results in an increase in the OH- ion concentration and increased soil pH. The extent to which exchangeable cations hydrolyse depends on their ability to compete with H+ ions for exchange sites. Ions such as Na+ are unable to compete as strongly as the more tightly held ions such as Ca2+ and Mg2+. For this reason exchangeable Na+2 and K+2 are hydrolysed to a much greater extent and produce a higher pH than do exchangeable Ca2+ or Mg2+. Hydrolysis of exchangeable Ca2+ and Mg2+ ions, in fact, is so limited that it results in a soil having only by a mildly alkaline reaction. Hydrolysis of compounds like CaCO3, and MgCO3, takes place according to the reaction:

$$CaCO_3 + 2H_2O \quad Ca^{2+} + 2OH^- + H_2CO_3$$

In this reaction H+ from water is inactivated through combination with carbonate to form weakly ionized carbonic acid. Hydroxyl ions are not inactivated through combination with Ca2+ resulting in an alkaline solution. The hydrolysis of CaCO3 and of MgCO3, is limited due to their low solubilities and therefore they tend to produce a pH in soils no higher than about 8.0 to 8.2. Soils containing measurable quantities of Na2CO3, have a pH of more than 8.2; the pH increases with increasing amounts of Na2CO3, and may be as high as 10.0 to 10.5. This is due to the higher solubility of Na2CO3 and therefore the greater potential for hydrolysis. According to Cruz-Romero and Coleman (1975) exchangeable sodium and CaCO3 react in low CO2 - low neutral salt environments to produce high pH and appreciable concentrations of Na2CO3. Since the soils of arid and semi-arid regions nearly always contain some calcium carbonate, a build up in the exchangeable sodium in the absence of an appreciable quantity of neutral soluble salts will always result in high pH; the exact value depending on the concentration of Na2CO3, formed or the level of ESP." [Ref: 7; Sodic Soils and Their Environmental management]

This is a common clay reaction that occurs in arid landscapes.

These conditions are typical of Australia's Great Artesian Basin, which has arid conditions throughout. At pH of 8.2 or more silica dissolves more readily but at pH of 10 and above Silica dissolves rapidly but in both the acid phase and the alkaline phase silica is dissolved as OH- ions are produced by both of these phases. The Acid sulfates and the neutral chlorides are well represented in the area of the Great Artesian Basin.

A quick analysis of these quotes shows us that there is indeed a spike in the pH but also on the way to producing this spike. Carbonic acid is formed, this would certainly dissolve some of the silica [Carbonic is a very efficient silica solvent] which at high pH would remain as colloidal silica [Silicic acid in solution]. The release of alkali ions such Na [sodium] also helps dissolve more silica. The OH- radical is also produced which is known to dissolve silica. The only ingredient now needed is Ammonia which is provided by decayed organic matter. Now conditions are set for opal to form.

This is a possible natural opal formation process which fits with the conditions and the natural resources of the GAB.

2.14 Physical Weathering

These are the mechanical or physical actions on the earths surface that result in the disintegration and wearing away of rock, or soil by agents such as wind, water or temperature changes. The actions of water include, frost, ice, rain, or snow.

This type of weathering mainly causes the formation of coarse soils and gravels. These include silts, sands and gravels. The Coarse products are derived from broken rock particles, whereas the sands and silts consist mainly of mineral grains.

2.15 Chemical Weathering

Chemical weathering produces mostly the very fine grain particles which make up the clays and clay - silt soils. Chemical weathering occurs in a wet and warm environment causing decomposition of many rock types to form fine grain soils. Few minerals are resistant to chemical weathering, Quartz being an exception. Chemical weathering can also cause alteration of some mineral species.

The main acids involved in chemical weathering are Sulfuric acid H2SO4 and Carbonic Acid H_2CO_3

[See Ch. 3. Sulfates; Carbonate Reactions]

Three main products of chemical weathering are:

- Haematite
- Micas
- Clay Minerals

Acid alteration in geological terms is called speleogenesis [CH. 2]

Sulfates & Hydrothermal Fluids

2.16 Water Salinity

A distinction is made between the salinity of inland waters to that of ocean waters, because of different salt contents. The following terms are widely used but not yet universally accepted; the term for the salinity of inland waters is saline whereas the term for the salinity of ocean waters is haline.

A number of other terms are used to distinguish the amount of salt contained in water. These terms are:

- Fresh water <0.05% salt or 500ppm
- Brackish water 0.05% - 3% salt, or 500 – 30,000ppm
- Saline water 3 – 5 % salt, or 30,000 – 50,000ppm
- Brine >5% salt, or > 50,000ppm

2.17 Salts & Brines

In the natural environment it is almost impossible to find samples that do not contain salts. Salts are everywhere. There are many different types of salts most of which are listed under evaporates, however this does not contain all the salts, there are also metal salts [oxides] and organic salts.

Environments where salts are found:

- All soils
- Hydrothermal waters
- Mineral waters

- River & sea water
- Volcanic fluids
- Lake Evaporites
- Organic matter

Salts are natural electrolytes. The Sulfates and Chlorides in particular are found widespread and in abundance. These salts are found in the Australia's Great Artesian Basin area in concentrations of approx. 3% strength. Most salts have an ammonium variety, which may be a key ingredient in opal forming.

"For the formation of precious opal Darragh et al. (352) point to three features that are important, at least in the Australian fields: (a) an abundant supply of readily soluble silica: (b) an arid climate restricting shallow groundwaters to sharply defined dams such as bentonite beds, which prevent the formation from drying out, thus retaining a typical solution containing up to 3% soluble sulfates and chlorides and 80ppm soluble silica: and (c) the presence of cavities, formed in various ways, in which silica particles can collect and arrange themselves (Ref: 22; Iler 399)".

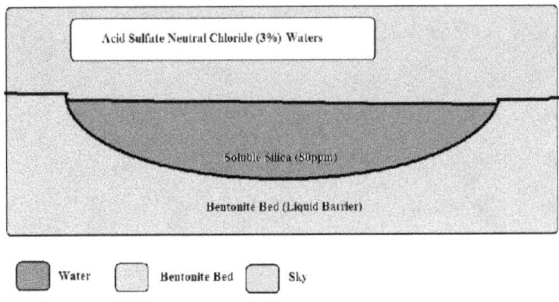

Figure 3 Acid sulfates & Neutral Chlorides

2.18 Brine

"A concentrated aqueous solutionof Sodium Chloride, in nature containing the cations Na+, Ca2+, K+ and Mg2+ but with Cl- the dominant anion, which is capable of LEACHING metals from rocks through which it passes." [Ref: 31; Philip Kearey; pgs 36, 109].

Natural brines are often called Geothermal or Hydrothermal brines. These brines can infiltrate and flow through sedimentary layers.

2.19 Electrolytes

Volcanic opal formation involves melted silica from a source such as Rhyolite which is hydrated [the silica sol takes on water] before the sol polymerizes to a gel and hardens to a solid.

Certain volcanic electrolytes act as catalysts and ion exchange mediums to facilitate the process of opal formation. Electrolytes can be liquids, solids or gases which conduct the passage of electrons through the electrolyte medium. Electrolytes are normally liquids. They can be any of the following:

- Acids
- Bases (Alkalis)
- Alcohols
- Water
- Salts
- Metals
- Some gases

The formation of opal requires an electrolyte for the ion-exchange process to occur. Silica can swap ions with some other elements. Opal contains some of these swapped ions, as impurities. Ion exchange may even be involved in the separation of silica from the clay, and the formation of silica spheres of the same size whilst rejecting silica particles which are too large for opal formation.

In nature there are many electrolytes. Determining the correct electrolyte for opal formation will probably mean finding the reaction that causes the formation in the first place. Electrolytes are often specific to a certain type of reaction. The type of reaction that we should be looking to find is one that produces a polymer or polymer product. Examples of Polymers are Glass and plastics.

Minerals, such as aluminosilicates, alumina, silica or zirconia can act as ion-exchangers because the skeleton or matrix material carries an excess charge which is neutralized by mobile counter ions. Anion - cation separations on a mixed-bed alumina - silica column. The metal oxide can act as either an acid or base and indicate the possibilities of cation- and anion-exchange behavior respectively:

There are 3 main types of electrolytes in the hydrothermal systems:

- Acid solutions, i.e. Acid Sulfate
- the Alkaline solution and the
- Brine or saline solution. The Saline solution is the one must be consider because NaCl is abundant in the opal environment (The Great Artesian Basin, was once an inland sea, thus large amounts of salt were deposited).

2.20 Sodium Chloride

Research has shown that in some hydrothermal waters the Sodium Chloride content can be as high as 50% of solution. This is far above the NaCl content of seawater, and such a saturated solution is referred to as being hypersaline. Sodium Chloride is a component of soils, minerals, clays, seawater, evaporites and groundwaters. It is the most prevalent of the chlorides.

I suggest that Silicification is to a great extent an ion exchange process with the electrolyte driving the process being firstly saline solution derived mostly from residual seawater which became enclosed by land as the Great Artesian Basin formed. As minerals began to leach into the solution it would have given way to the acid sulfate process.

It has already been noted that Potassium Feldspar provides one alkali ion that helps in the dissolution of silica. Other alkali ions are present in the groundwater brine notably the sodium ion and also the chloride ion.

The chlorides probably help to release silica from the clays by combining with the metal ions in the clays thus acting as chelating agents stripping the clays of their metal ions. It is not just the chlorides that act this way but the sulfates, especially the acid-sulfates which also may act as chelating agents.

The sodium ions along with potassium from the feldspar then attack the silica, dissolving it. NaCl is an ionic polyelectrolyte; polyelectrolytes are known to cause gelling in silica, although there are many other polyelectrolytes. Gelling is the result or product of polymerization.

The question that remains then is where do these salts come from and are they responsible for the alteration of K-Al-Si to precious opal. Some sources state that NaCl may inhibit silica polymerization. This may be dependant on the concentration and interaction of salts, and their ability to decompose one another which could endow them with ion exchange capabilities.

The fluids within the sedimentary layers are called hydrothermal fluids. Hydrothermal simply means heated water. These fluids however are influenced by their environment and the mineral species contained therein. Hydrothermal fluids vary greatly in composition, however the main alteration types are:

- Chloride (Acid Chloride)
- Acid-sulfate

Both chlorides and sulfates are classed as salts, however in the acid form they are much more reactive and produce numerous byproducts.

All these salts come from weathered mineral products and from their immediate sedimentary environment. Minerals are leached from the sediments by the acids and water. As the minerals are dissolved, their ions are free to recombine with other elements; this is the process of alteration mineralogy.

2.21 Effect of pH

The electrolyte should be at pH 7 + (for alkaline solution) as this is the required pH for silica to dissolve to produce Silicic acid which is released into the solution. In order to be confident that the required pH has been reached, you will need a way to measure the pH of the electrolyte solution. The best way is to use a pH meter.

The correct pH for an acid solution however will be different and at the lower end of the scale, possibly about pH 3-4, or even as low as 2.

The importance of achieving the right pH cannot be underestimated because without the release of Silicic acid, you have no hope of polymerization; pH is also required for polymerization itself. (The K-Feldspar may not be enough to reach the required pH level).

The electrolyte should then be maintained at a specific pH. The pH of many reactions are controlled by:

TABLE 2: pH Controllers:	
Most Bases [Alkaline] Adjust pH **up**	Most Acids Adjust pH **Down**
Sodium Hydroxide [caustic soda]	Sulfuric Acid
Calcium Hydroxide	Hydrochloric Acid HCL
Potassium Hydroxide	Phosphoric Acid
Sodium Carbonate [Washing powder]	Alum [Aluminium Sulfate]
Bicarbonate Soda	Trichlor [pool chemical]
Sodium Hypochlorite	Dichlor [pool chemical]
Calcium Hypochlorite	Chlorine gas
Algal growth	Organic litter

Swimmers wastes	Rain water

Besides the pH controller you will also need a Silicic acid polymerization initiator. A known gelator for Silicic acid is Aluminium Hydroxide $Al_2(OH)$. (See Ch.4 Polymerization).

Proof that the electrolyte is saline is the fact that the silcrete that caps natural opal formations is formed in a saline environment. This is an integral part of the opal formation process and in fact is a byproduct of it, this is all the material that could not be converted to opal.

It is already known that in the K-Al-Si system that a Chloride electrolyte will act to release silica (silicification process) and that NaCl is a polyelectrolyte which will cause polymerization of silica. However the same may be achieved by acid-sulfate.(see Ch. 4. Polymerization).

2.22 Acid-sulfates

(Include $NaHSO_4$ Bisulfate..Acid Chlorine Sulfates $ClSO_4$)

Acid-sulfates are the product of the hydrolysis of the Sulfides. The most common of the sulfides by far is Iron Sulfide (FeS_2). It is contained in all the major rock types. There is an abundance of FeS_2 in Hydrothermal solutions, in sedimentary deposits and in some types of soils. The sulfuric acid that is released combines with many different metals to form metal sulfates.

In the mining environment FeS_2 is a waste product and is quite problematic because of its high reactivity and high concentration. When rain comes acid can be released by hydrolysis from mine tailings into the environment where it becomes a toxic problem to plants and animals.

What are the chemical clues that will lead us to the Opal electrolyte.

From researching different hydrothermal fluids and the resulting alteration mineralogy, and gangue mineralogy, I have decided that the electrolyte could be acid-sulfate. Further investigation needed.

This decision was influenced by observing the associated mineral alteration and salts that are associated with Opal formation. FeS_2 was a component of all the hydrothermal fluids that I evaluated that they were associated with opal formation (Only some hydrothermal fluids are associated with Opal). It was also present in the sediment associated with Opal. FeS2 is known to release carbonic acid from carbonates. Carbonates are often associated with sandstone acting as a binding agent.

There are two main alteration products that are nearly always associated with Opal, these are:

- Alunite $KAl_3[(OH)_6][(SO_4)_2]$
- Glauberite $NaSO_4.CaSO_4nH_2O$

Alunite is formed by the action of FeS_2 on Kaolinite, which is a part of the opal dirt formula. It is the released Sulfuric acid H_2SO_4 that actually causes the alteration. The Alunite may act as the polymerization initiator, which begins the process, as Aluminium hydroxide is a well known inorganic gelator.

The Glauberite is a byproduct of the action of the released sulfuric acid from the FeS_2. The Glauberite may help in the dehydration process, although evaporation is probably the main contributor to dehydration.

Gangue mineralogy must also be considered as a clue to the formation of opal. You must understand that different alteration types will contain different gangue minerals. The main relevant gangue minerals associated with Australian sedimentary opal formation are:

- Gypsum $CaSO_4$
- Alunite $KAl_3[(OH)_6][(SO_4)_2]$
- Hematite (Fe_2O_3),
- Limonite $FeO(OH) \cdot nH_2O + Fe_2O_3 \cdot nH_2O$ (Iron Hydroxide + Iron Oxide + water)
- Geothite & Limonite $= HFeO_2$ (Iron Oxyhydroxide)
- Celadonite $K(Mg,Fe++)(Fe+++,Al)[Si_4O_{10}](OH)_2$
- Sulphosalts
- Pyrite (FeS_2)

Celadonite is found in opal filled cavities, which suggests that it could be the initiator or at least an indicator for opal polymerization. It has been dismissed however, of any connection with opal by geologists due to the fact that the unfilled cavities are also coated with Celadonite.

Note: Iron oxides and hydroxides are byproducts of FeS_2 hydrolysis reactions (as seen below, see reaction formula for FeS_2 Hydrolysis).

Some zeolites may also be present.

Celadonite is a phylosillicate

2.23 Volcanic Acid Sulfates

At first it was hard to see the relationship between Acid sulfates and opal formation, but the natural evidence is overwhelming. Volcanic opal forms from Acid sulfate fluids where they mix with neutral chlorides. There is evidence that sedimentary opal of Australia also form in an Acid sulfate and chloride electrolyte.

This made me ask the question "What reaction does acid sulfates & Chlorides have with clays?" What I found was that the Acid sulfates & Chlorides did not dissolve the silica but it did dissolve the Alumina thus breaking the aluminosilicate bond which allows the silica to dissolve more easily. Refer figure 3.

It is the Al ions in the soil that drive the pH scale to very low levels as most of the acidity in soils is caused by the Al3+ ion. For every mole of Al3+ resulting from hydration it produces 3 moles of acidity (H+).

2.24 Sulfuric Acid H_2SO_4 Alteration

The By-products which are related to H_2S-H_2SO_4 materials that influenced speleogenesis of Carlsbad, Lechuguilla and other caves of the Guadalupe Mountains in New Mexico.

"Carlsbad Cavern, Lechuguilla Cave, and other large caves of the Guadalupe Mountains, New Mexico, contain minerals and amorphous materials derived as by-products from H2SO4 speleogenesis (process of cave formation influenced by sulfuric acid). These materials, referred to as "speleogenetic by-products," are categorized as primary or secondary. Primary speleogenetic by-products are formed directly from H2SO4 speleogenesis by H2S-H2SO4 reaction with carbonate bedrock and internal sediments. They are found in cave areas protected from flood or drip waters. Secondary speleogenetic by-products are minerals and amorphous materials that formed by the alteration of the primary speleogenetic by-products, or by the late-stage remobilization of elements concentrated during speleogenesis. Primary speleogenetic by-products in these caves are gypsum, elemental sulfur, hydrated halloysite, alunite, natroalunite, jarosite, hydrobasaluminite, quartz, todorokite, rancieite, and amorphous silica and aluminum sulfates. Gypsum is the most abundant by-product, whereas alunite is the most significant because it can reveal the timing of speleogenesis. Aluminite, tyuyamunite, quartz, opal, and gypsum are secondary speleogenetic by-products. Other possible speleogenetic by-products are celestite, hydrous iron sulfates, gibbsite, nordstrandite, goethite and dolomite. The carbonate bedrocks in which the caves have formed are predominantly dolostone and limestone; mineral assemblages of these host rocks include calcite, dolomite, quartz, illite, dickite, kaolinite, interstratified illite/smectite, montmorillonite, and mica. The process of H2SO4 speleogenesis for Carlsbad Cavern and Lechuguilla Cave, and its by-products provide a general model for similar cave and non-cave systems worldwide (Ref: 9; Jim Pisarowicz).

It seems from the above quote that sulfuric acid has a major alteration affect on many minerals including the formation of opal. The major source of sulfuric acid is usually FeS_2 Iron

Pyrite.

2.25 Organic Formation of FeS$_2$

"As sulfates dissolve in salty water which is converted into hydrogen sulfide that then react with available iron oxides in sediments to form Iron sulfides. When these iron sulfides are exposed to the air they then release sulfuric acid. (Acid-Sulfate connection to Salt Brines).

The chemical reaction is as follows:

Iron oxide (sediment) + Sulfate (seawater) + organic material ➜

Iron sulfide (pyrite) + carbon dioxide + water

The organic material would normally be from rotting vegetation; this reaction occurs in a low oxygen environment. The organic process is aided by bacteria.

FeS2 hydrolysis

Iron pyrite + oxygen + water ➜ *sulfuric acid + iron hydroxide + iron sulfate*

$$FeS2 \quad + \quad O2 \quad + H2O \rightarrow H2SO4$$
$$+ FeOH \quad \quad + FeSO4$$

Hydrogen Sulfide (H2S) is a product of anaerobic breakdown of organic materials. During manure storage, sulfate is reduced to hydrogen sulfide" (Riviere, 1974).

"Sulfates are indirectly responsible for odor and corrosion of waste handling systems resulting from reduction of sulfates to hydrogen sulfide under anaerobic conditions (Sawyer and McCarty, 1978)".

Anaerobic
$SO2-$ + organic matter ----------> $S2-$ + $H2O$ + $CO2$
bacteria

$S2-$ + $2H+$ ----------> $H2S$

(Ref: 10; David Schmidt)

It should also be possible even probable to produce Hydrogen Sulfide from chicken manure, (or other decomposing organic matter) as it was a major problem for the mushroom farm in their composting of straw with chicken manure and water. The odour of H_2S (hydrogen sulfide) was very strong. The chicken manure may have to be sourced from a farm.

Sulfuric acid falls in rain and reacts to corrode iron.

$$H_2SO_4 + 4Fe \rightarrow 4FeO + H_2S$$

The action of sulfuric acid on metallic iron (Fe)

2.26 Chelation

If other chlorides are required then these can be added to the solution in the form of chlorine (swimming pool chlorine) which will combine with the metals in the clay forming metal chlorides, and once again stripping the clays of their metal ions allowing the silica to dissolve more easily.

Recent Information from Sudan

"A recent hydrothermal mud pool at the southwestern slope of the Rincon de la Vieja volcano in Northwest Costa Rica exhibits an argillic alteration system formed by intense interaction of sulfuric acidic fluids with wall rock materials. Detailed mineralogical analysis revealed an assemblage with kaolinite, alunite, and opal-C as the major mineral phases. Electron paramagnetic resonance spectroscopy (EPR) showed 3 different redox-sensitive cations associated with the mineral phases, $Cu\{sup+\}$ is structure-bound in opal-C, whereas $VO\{sup\ 2+\}$ and $Fe\{sup\ 3+\}$ are located in the kaolinite structure. The location of the redox-sensitive cations in different minerals of the assemblage is indicative of different chemical conditions

The formation of the alteration products can be described schematically as a 2-step process. In a first step alunite and opal-C were precipitated in a fluid with slightly reducing conditions and low chloride availability. The second step is characterized by a decrease in $K\{sup+\}$ activity and subsequent formation of kaolinite under weakly oxidizing to oxidizing redox conditions as indicated by structure-bound $VO\{sup\ 2+\}$ and $Fe\{sup\ 3+\}$. The detection of paramagnetic trace elements structure-bound in mineral phases by EPR provide direct information about the prevailing redox conditions during alteration and can, therefore, be used as additional insight into the genesis of the hydrothermal, near-surface system". (Ref: 15; Dr. Mario Wipki;)

2.27 Composition of Opalising Fluids

"There has been much debate in recent times concerning the composition of Opalising fluids in the formation of opal in Australia. In most other nations the Opal is of a volcanic origin and obviously form in volcanic fluids. It is known that the hydrothermal alteration of felsic volcanics by acid sulfate fluids and from the boiling and/or mixing of neutral chloride geothermal brines in epithermal settings (Corbett & Leach 1998)"

(Ref: 21; Regolith 2004, pg 264).

Despite the fact that the Australian deposits are considered to be of sedimentary origin and form at temperatures below 100C, the Acid Sulfates and Chlorides are present in the opal environment at the concentration of approx. 3% (Ref: 22; Iler, 399).

2.28 Explaining the Sulfuric acid link

Sulfuric acid is indicated by the US Geological Survey [ref: 8] as possibly being responsible for opal formation. This indication is by individual analysis of the associated minerals in the sedimentary opal environment. All the associated minerals are sulfates or iron derivatives of FeS2. Carbonates and FeS2 are widely available throughout the opal environment.

Others suggest Speleogenesis, a form of sulfuric acid alteration as a cause of opal formation. This does include Carbonic acid alteration also and is used mainly to describe the formation of caves, in which opal sometimes forms.

Volcanic opal formation has been linked to the mixing of Acid
Sulfate fluids with neutral chlorides. This results in the
neutralization of the sulfuric acid with the production of
sulfates, and iron species. The neutral chlorides will produce
some HCL until they are neutralised by bases such as
carbonates. This tends to produce Sodium carbonates and
possibly some calcium chloride or even iron chloride which
quickly raises the pH of the mixture to a pH above 7. Iron
Chloride is known to etch silica.

The Australian opal fields contain abundant sulfates and
chlorides to about 3%. Much of the Australian mainland has
layers of Iron Sulfates these tend to oxidize releasing
concentrated sulfuric acid and Fe_2+ ions and Fe_3+ ions. These
ion species help to breakdown organic matter and can produce
the Fenton reaction in water. The Fenton reaction is also
known to breakdown organic matter.

The evidence then that sulfuric acid is involved in opal
formation is strong. Sulfuric acid however is a poor solvent for
silica.

Sulfuric acid action on carbonates dissolves silica but requires a
solvent such as an alcohol or an alkane to prevent immediate
gelling. Once gelling has occurred the chance of particle
growth is slim. This reaction produces Carbonic acid which is
responsible for silica dissolution in natural chemical weathering
of soils and rock.

As sulfuric acid reacts with carbonates much foaming occurs
and dissolved silica will overflow the container if container
does not have a long necked.

This may be a viable way of producing SiO_2(aq) for opal

production as the other opal precursors only achieve the same after polycondensation.

This method will achieve dissolved silica quickly as carbonic acid rapidly dissolves silica, however is an exothermic reaction and will produce boiling temperature [100°C]. Any protein & enzymes should be added after this solution cools a little. Once this is achieved then it will need to be aged at pH 7 and heating above 35 but below 60°C.

Sulfuric acid readily bonds with alkanes [Mineral oils] to form Sulfonic acid an ion exchange medium however this is a reversible reaction and the sulfonic acid reverts to sulfuric acid again.

Siloxane bonds form in silica at low pH but silica does not readily form bonds with hydrocarbons. It is possible that the sulfonic acid exchanges the alkane ions with the silica, forming something like dimethyl siloxane.

Polysiloxanes such as dimethyl siloxane are very good ion exchange mediums. Polysiloxanes are known to be condensed by Hydrolytic enzymes, once again forming SiO_2 (aq) but having initiated ion exchange.

The alkanes such as the mineral oils result from the breakdown of organic matter. Some scientists and geologists have long thought that there is a link between the formation of crude oil and the formation of opal. The alkanes could well be this link.

The formation of polysiloxanes by the above method is put forward by the author as a suggested way that they could form in nature. It is not yet verified, but looks very plausible.

NOTE: If you use the Protein/Hydrolytic Enzyme system you won't need Ammonia(aq) as the breakdown of the protein, and the amino acids will provide both some ammonia gas which will become aqueous in water and amines, which can be substituted for ammonia.

Using this method you will not be handling Ammonia or amines as these will be produced in your glass reaction vessel. Len cram uses Glass jars, such as jam jars or vegemite jars. MALODOURS may be produced.

2.29 Search for the Opal ion exchanger

In order to measure the strength of an electrolyte the Cation Exchange Capacity (CEC) has been developed as a standard for measuring. CEC is also a measure of the electrolytes ability to complex metals (Chelation). The electrolyte involved in the transformation of silica to opal must be a strong one. Let's have a look at some electrolytes and their CEC's.

Sulfuric acid is definitely a strong electrolyte but doubt exists as to its direct involvement in the formation of opal. Sulfuric acid alteration has been studied for many years without a solid direct link being established to the formation of opal. Although many sources point to sulfuric acid as the alteration electrolyte, it has never been proven.

In fact the hydrothermal alteration electrolytes need to have carbonates present to form carbonic acid before large amounts of silica will dissolve. The ability of Sulfuric acid to dissolve silica at ambient temperatures seems to be very limited without the aid of carbonates.

It is interesting that Sulfuric acid reacts with carbonates to release Carbonic acid. Carbonic acid on the other hand successfully dissolves silica at ambient temperatures; we know this from chemical weathering studies. Thus Carbonic acid must be considered along with the Acid sulfates and Chlorides in order to establish the electrolyte [Opalising fluids] that initiates the Opal formation process. Phosphates/Phosphoric acid also a possibility – an organic acid that dissolves some quartz types unheated.

However, the main ion exchanger in the opal process has been identified by Byron Deveson, as clay called Montmorillonite. It displays extraordinary ability to exchange ions even to the point of producing ultra pure water in a sedimentary environment. [More info under Mound springs].

2.30 Hypochlorites & FeCL

It is highly likely that Acid sulfates when they mix with neutral chlorides may form opal. This particular mix of salts has the special ability to be able to decompose one another. This decomposition may cause the formation of other electrolytes, possibly thiosulfates, which have been found in hydrothermal fluids, possibly even forming ammonium bicarbonate. It has to be remembered that the term **'acid'** as used by geologists when referring to rocks is often a reference to silica content which is present as Silicic acid.

Thiosulfates have been found in hydrothermal waters and are a possible product of the Acid Sulfate and Neutral Chloride mixing as sometimes occurs in nature in volcanic areas.

Thiosulfates are also used in the photography industry to develop film. Their affect if any on the opal forming reaction is as yet unknown.

FeS_2 although it is Iron Sulfide, is often referred to as an acid sulfate, it is probably THE acid sulfate. Assume for the neutral chloride NH_4CL, Ammonium Chloride. In this reaction, the chlorine ion neutralizes the acid sulfate producing Iron Sulfate and Iron Chloride. Alkalis ions present in the mix react with the chlorine ion to form hypochlorites.

The hypochlorite family consists of:

- o NaCLO: Sodium Hypochlorite
- o CaOCL$_2$: Calcium Hypochlorite
- o HOCL: Hypochlorous acid
- o HCLO$_4$: Perchloric Acid [Strong Acid]
- o HCLO$_3$: Chloric Acid
- o HCLO$_2$: Chlorous Acid
- o HCL: Hydrochloric acid

The properties of this group of chemicals is that they are mostly used as bleaches and are very strong oxidizing agents and can be very explosive in the wrong environment. How this reaction interacts with the Silica is as yet unknown. Most of this chemical family degrades quickly, therefore does not persist in solution for extended periods of time, some only need sunlight to begin the decomposition process. Some type of Fenton reaction may be taking place simultaneously. Normally this reaction uses the catalyst Hydrogen Peroxide as an oxidizing agent but could easily be replaced by a hypochlorite.

Acid Sulfate/Chloride waters tend to be a green colour due to the dissolved Iron content. This same green sulfate/chloride water is found within opal producing areas. The hypochlorites formed by the acid sulfate chloride reaction are all short lived, dissipating into non reactive chemicals, apart from the FeCL which would remain as an active electrolyte for an extended period of time. Another possible way that FeCL could form is as a result of the Fenton reaction.

Figure 4: Acid Sulfate & Neutral chloride reaction

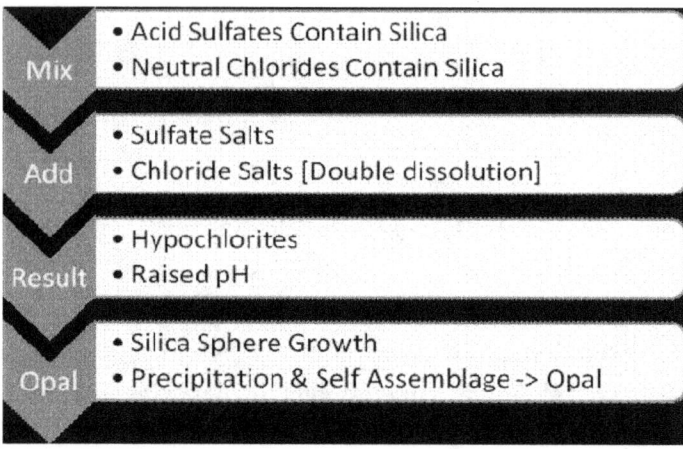

Mix
- Acid Sulfates Contain Silica
- Neutral Chlorides Contain Silica

Add
- Sulfate Salts
- Chloride Salts [Double dissolution]

Result
- Hypochlorites
- Raised pH

Opal
- Silica Sphere Growth
- Precipitation & Self Assemblage -> Opal

It has been quite obvious that Sulfuric acid does form through inorganic chemical reactions. However Sulfuric acid also forms from organic matter decay.

Sulfuric Acid Carbonate Reaction:

$$H_2SO_4 + CaCO_3 [H_2O] => Ca_2SO_4 + H_2CO_3$$

Ca_2SO_4 = Calcium Sulfate [Clay Dispersant]

H_2CO_3 = Carbonic acid [Clay Solvent]

Unlike most other "so called" clay solvents, carbonic acid actually dissolves the silica and alumina, silica gel is produce very efficiently however often becomes a hard gel, rather than a soft gel which is needed for a colloid to form. A slower reaction, which produces less heat [exothermal] may produce a soft gel which will be hydrophilic, allowing a colloid to form.

Carbonic acid is also formed in the breakdown of Ethanol:

Ethanol => Acetic acid => Carbonic acid

Acid sulfates definitely seem to be responsible for the breakdown of organic matter into its final end products which then interact with the silica at elevated pH conditions, conditions which are conducive to the dissolution of silica, forming silica colloids. At pH above 9 the ammonium ion from the breakdown of the organic matter becomes Ammonia gas, or possibly aqueous ammonia, bubbling up through the mixture, catalyzing the growth of silica spheres [Ostwald ripening].

[See Ch 3. Organic Decay; Ref 3:24.]

Carbonate Reactions

TABLE 3: Soluble & Insoluble Carbonates [Ash]	
Soluble Carbonates	**Insoluble Carbonates**
potassium carbonate K2CO3	zinc carbonate
magnesium carbonate	calcium carbonate CaCO3
sodium carbonate and Na2CO3	lead carbonate
ammonium carbonate [Urea]	copper (II) carbonate

2.32 The Role of Carbonates

Carbonic acid is also formed by the breakdown of organic matter.

Acid sulfates then directly and indirectly contribute to the formation of carbonic acid.

There are certain carbonate reactions that must be investigated in the study of opal formation for two distinct reasons:

- 1. They are part of the sedimentary opal environment being a major sediment of the Great Artesian Basin.
- 2. They have also been implicated in speleogenesis

All carbonates and hydrogen carbonates react with acids to form salt, water and carbon dioxide.

Dolomite is a mixture of Calcium & Magnesium carbonates.

Limestone is mostly Calcium Carbonate.

2.33 The Chemistry of Opal Formation

<u>Speleogenesis:</u> Alteration products of certain acids, related to cave formation; and the formation of many different minerals. The two main

acids involved in speleogenesis are: Sulfuric acid H_2SO_4 & Carbonic Acid H_2CO_3.

FeS_2 and the Fenton reaction may also be involved; as the Fenton reaction breaks down organic matter.

Carbonic Acid & Dissolved Silica

The chemical weathering of clays is mainly due to Carbonic acid; which dissolves silicates and breaks-down clays by dissolving the silica. The freed Aluminium then becomes hydrated due to the water content and will form Aluminium Hydroxide Al(OH) 3. In strongly acid soils which are especially associated with the humid tropic zones, Al plays an important role in soil buffering. The dissolution of silica by Carbonic acid releases Al3+ which is precipitated as Al Hydroxide by the following reaction:

$$Al3^+ + H2O? \ Al(OH)2+ \ + H+$$
$$Al(OH)2 + H2O? \ Al(OH)2+ \ II+$$
$$Al(OH)2 + H2O? \ Al(OH)3 + H+$$

Figure 5 Alumina Hydration [Ref: 37]

2.34 Carbonic Acid as Electrolyte

Carbonic acid may be formed by a number of different natural processes, one of the simplest ways for it to form is by CO_2 being released into H_2O water. It takes a hydrogen ion from the water and also another oxygen making H_2CO_3. This also has the effect of causing the water to behave as an hydroxide due to the fact that the water has lost a hydrogen ion, thus the water is now OH rather than HOH, this is referred to as a Reactive Oxygen Species (ROS). It is this tendency to split water into H^+ ions and OH⁻ Hydroxide ions, that makes Carbonic acid a good electrolyte, it is rather the ROS than the carbonic acid that is the good electrolyte. This helps to explain Carbonic acids effectiveness in dissolving Silicates, and silica from clays. [See Annex B → Reaction Formulas: Reactive Oxygen Species].

The dissolved silica (Silicic acid) H_2SiO_4 polymerizes with water to form

$SiO_2 + H_2O$ which is a reversible reaction but in the presence of Ammonia NH_3 it forms silica spheres which gell. The gel would remain soft if it were not for the presence of $Al(OH)_3$.

As the solution begins to dry out; dehydrate; the opal gel also begins to harden, catalyzed by the formation of Al2O$_3$; Aluminium oxide formed by the dehydration of Al(OH)$_3$.

Draining the liquid will accelerate this process; which could still take months to complete even in arid conditions, a little extra heat may help this process.

2.35 Salts of Carbonic Acid:

- Potassium Carbonate K_2CO_3
- Sodium Carbonate Na_2CO_3
- Urea [Ammonium Carbonate]

Bicarbonates, also called acid carbonates or hydrogen carbonates.

Carbonic Acid can be formed by hydrolysis or acid hydrolysis of carbonates. However Carbonic acid is unstable and quickly decomposes forming CO2 and H2O. This could possibly cycle to form the acid again. However there are many sources of CO_2 which are:

- o Byproduct of Fermentation
- o Atmospheric CO_2 transported by rain, producing H_2CO_3
- o Plant roots release CO_2
- o Microbes release CO_2
- o Decomposition of plant & animal matter
  - ▪ Glycolic Acid
  - ▪ Oxidizing Coal with Alkaline

Permanganates
- Carboxylic Acids
o Acids dissolving Carbonates
o Water bourne Carbonates and Bicarbonates

2.36 Hydrating Agents

Carbonic acid is known to be a hydrating agent. This may answer the question of how water gets into the opal structure, hydrating agents facilitate water retention.

Many of the Alpha Hydroxy Amino acids (AHA's) are also hydrating agents and may also be involved in this process of hydration of Silicic acid as it forms into opal gel.

See also: Dehydrating agents & Absorbents

Carbonic Acid is made commercially, this will be concentrated acid i.e. no water.

Carbonic acid formed by the action of sulfuric acid on carbonates. Thus Calcium carbonate plus sulfuric acid without water should yield:

$$CaCO_3 + H_2SO_4 \rightarrow H2CO_3 + CaSO_4$$

Concentrated Carbonic acid and Calcium Sulfate.

If Sodium Carbonate with water was also present, it should result in the formation of the carbonic acid and Glauberite

$[NaSO_4.CaSO_{4x}H_2O]$.

Glauberite often found with opal, could well be a byproduct of the opal forming process, as is Calcium Sulfate [Gypsum].

2.37 Some Organic Reactions

We see Carbonic Acid being produced in nature by both Inorganic and Organic chemistry. These methods are:

- Inorganic – Action Sulfuric acid on Carbonates [speleogenesis]
- Organic – Degrading of Carboxylic acids [organic decay]

As Carbonic acid dissolves silica from aluminosilicates it releases Al into solution as Al_3^+ ions which hydrate to Aluminium hydroxide $[Al(OH)_3]$ along with the other main products of decay;

- Ammonia
- Phosphoric acid

Now let's have a quick look at the function of these chemicals in the Opal formation process.

Functions:

1. Carbonic acid; dissolves silica; releases Al_3^+; Hydrating agent.
2. Silicic acid [Silica acid]: with water forms opal gel
3. Aluminium [forms $Al(OH)_3$] Silica Polymerizing agent; dehydrates to Aluminium oxide $[Al_2O_3]$ which dehydrates the opal gel.
4. Ammonia or Amines/Polyamines [Silica sphere control]
5. Phosphoric acid; another Silica Polymerizing agent
6. Urea degrades to Ammonia & Carbonic acid [CH 3. Ref:17]

2.38 The Role of Iron

Iron accelerates the decay of organic matter, this is especially the case in the Fenton reaction. The Diiron species may not be requires for the polymerization of silica gel as two other polymerizing agents Aluminium hydroxide and Phosphoric acid are produced in the opal forming process.

The role of the iron may simply be to accelerate the decay of organic material.

2.39 More Carbonic Acid

Decay of plant and animal matter produces an abundance of Carbonic acid in many and varied chemical reactions, some have been listed above but more are described in the following few pages which simply highlights the plentiful supply of carbonic acid, some by inorganic reactions and some by the natural decay process.

Carbonation drops may be used to form Carbonic acid; or Sodium bi-carb with acetic acid; or FeS_2 + Carbonates; or from Carboxylic acids.

Not only is Carbonic Acid a decomposition agent, it is one of the main end products of decomposition, as all organic matter breaks down to Carbon, the form of Carbon is Carbonic acid. That is to say that carboxylic acids which contain the COOH group break down to Carbonic acid.

Decomposition End Products of Amino Acids

• Carbonic acid -decomposition of Carboxylic acids

- Ammonia - from decomposition Amines
- Nitric Acid - "   "   "
- Ethanol--> from fermentation + decarboxylation
- CO_2 to form more Carbonic acid + fermentation.

2.40 Carbonic anhydrase. It is found in red blood cells where it catalyzes the reaction

$$CO_2 + H_2O <-> H_2CO_3$$

It enables red blood cells to transport carbon dioxide from the tissues to the lungs.

One molecule of carbonic anhydrase can process one million molecules of CO_2 each second.

However one KEY is that an enzyme Carbonic Anhydrase catalyzes the binding of CO_2 and water (H) to formation of carbonic acid. This is of GREAT significance because under normal circumstances only 1% of the CO_2 in water becomes Carbonic Acid.

This is a SPECIES SPECIFIC reaction as are most Amino Acid Enzyme reactions.

2.41 Carbonation

This occurs when carbon dioxide is dissolved in water or an aqueous solution. This process is generally represented by the following reaction, where water and gaseous carbon dioxide react to form a dilute solution of carbonic acid.

$$H_2O + CO_2 \leftrightarrow H_2CO_3$$

This process yields the "fizz" to carbonated water, the head to beer, and the cork pop and bubbles to champagne and sparkling wine. Carbonation is used to improve both the taste and "texture" of the carbonated consumable. Carbonation is sometimes used for reasons other than consumption, to lower the pH (raise the hydrogen ion concentration) of a water solution, for example.

2.42 Amino Acids with Carbonic Acid

The Amino acids provide both the Amines (NH_2), and CO_2 for the carbonic acid along with Carbonic anhydrase in water at low temperatures (up to 60°C) so that the enzymes remain active. The Carbonic acid dissolves the silica, hydrates the Al_3^+ to Al_2O_3. It also helps to hydrate the Silicic acid, along with the ROS forming Opal Gel.

2.43 High Temp Water Carbonic acid formation

Urea provides the NH3 by hydrolysis and the High Temp Water (HTW). It is an efficient medium for producing Carbonic acid. At HTW there are naturally more H+ ions and OH- hydroxide ions simply due to the water temperature, at these temperatures CO_2 combines more readily with the H+ and OH- ions to form carbonic acid. This is an efficient method of producing H_2CO_3. The source of CO_2 may well be from UREA.

2.44 Alumina Hydrating agent

Copper metal (powder) reacts with Al_3^+ in water; which should oxidize the Al_3^+ to Al_2O_3 and the copper also oxidizes. This releases the silica which may now dissolve to Silicic acid needed to produce opal. A source of Ammonia will still be required.

Too much SALT may interfere with the sphere formation of the Silicic acid, causing the spheres to be irregular and therefore will not pack properly into an ordered array. This will cause potch to form rather than precious opal.

2.45 Carbonates inhibit Fenton reaction

Despite the fact, that carbonates inhibit the Fenton reaction when they come in contact with H_2SO_4 they produce carbonic acid which dissolves silica and catalyses the hydration of Al_3^+ to Aluminium Hydroxide in water. The ammonium ion also dissolves alumina, leaching it from clays, causing it to hydrate to produce the gel Aluminium hydroxide which can be a white or brown gel which is a silica polymerization initiator.

When Acid Sulfates mix with Neutral Chlorides

2.46 Mixed Salt Reactions

Studies of the formation of volcanic opal have shown that the volcanic fluids that produce opal are Acid Sulfates where they mix with neutral chlorides. This would result in the following reaction:

Sulfuric acid with Sodium Chloride:

$$NaCl + H_2SO_4 \rightarrow NaHSO_4 + HCl\uparrow$$

$$2\,NaCl + H_2SO_4 \rightarrow 2\,HCl + Na_2SO_4$$

This may be an exothermic reaction and may also produce HCL gas.

HCL an acid is very reactive and will neutralize with bases [Alkaline ions] to form Chloride salts, with NaCl being the most abundant.

Salt [NaCl] needs to be converted by the addition of $CaCO_3$ to form Na_2CO_3. [See: Ref: 24]

Acid Sulfates fluids that mix with neutral chlorides have been identified as opal forming volcanic fluids. If these fluids come into contact with $CaCO_3$ then Na_2CO_3 will form changing the pH from Acid to Alkaline normally within the pH range of 8.3

– 10.5.

By far the most abundant of the neutral chlorides is NaCl [Sodium Chloride]. When Acid Sulfates mix with NaCl then the following reaction occurs:

The reaction of Sodium Sulfuric with Calcium Carbonate is:

Na_2CO_3 Sodium Carbonate [Washing Powder]

The remaining HCL tends to be very reactive and form other neutral chlorides and the sulfate ions will form other sulfates.

The effect of the Na_2CO_3 is to greatly increase the pH to levels between 8.3 and 10.5 but can be higher depending on the amount of Na_2CO_3 produced.

At raised pH colloidal silica is produced which is of paramount importance in the formation of opal. Colloidal silica SOLS forms in the pH range 7-10 in the absence of salt. The presence of salt causes Gelation, as does Silica saturation. NaCl can be neutralized with Calcium carbonate.

Where carbonates are present it would not be unusual for some carbonic acid to form also, which can rapidly dissolve silica.

Some studies into volcanic opal formation suggest that opal is formed where Acid Sulfates mix with neutral chlorides. Therefore a typical reaction would be:

ACID SULFATE:

$$FeS2 + H2O \rightarrow Fe2+ Fe3+ H2SO4 \rightarrow$$

NEUTRAL CHLORIDE:

$$NaCL \rightarrow FeCL + FeSO4 + FeOH + NaSO4 + HCL$$

The main host rock which these fluids act upon to create opal is thought to be Rhyolite, which is very similar in composition to Volcanic ash. [Ref: 21; pg 264]

2.47 Opal genesis by mixing of Acid Sulfates and Neutral Chlorides

In volcanic terms when referring to rocks as being acid is often used as a synonym for rocks containing a high silica content [greater than 63% SiO_2 by weight], such as the volcanic rocks, Rhyolite & Dacite.

[Ref: 32]. It is not hard to take this analogy a little further and use it in reference to Acid Sulfate the volcanic fluids from which these rocks originate. The main acid could well be Silicic acid rather than a sulfate acid which is probably only secondary by volume to the silica acid.

The other volcanic fluid mentioned as mixing with the acid sulfate to form opal is the neutral chloride. What are the neutral chlorides? These are salts that contain any of the ions listed under *ions of neutral salts* in the table below. However by far the most common is NaCl [Ref: 33].

[More information on salt reactions (K-Al-Si systems in Chloride electrolytes) and their solubilizing affect on silica see Ref: 35 & 36].

Figure 6: pH of Salts in water [Refer: Hydrolysis – acidic, basic, and neutral salts; pg 1 of 5

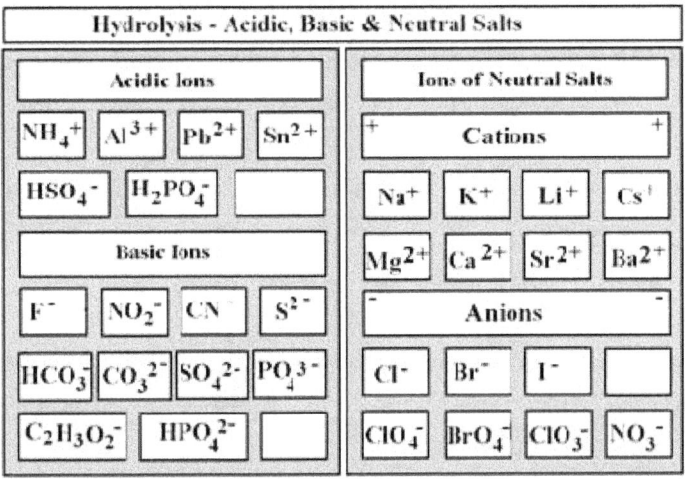

http://www.science.uwaterloo.ca/~cchieh/cact/c123/salts.html]

Soluble silica as the opal source is melted by volcanic heat before meeting and mixing with the neutral chlorides. The exact mechanism by which this volcanic opal forms is still unknown.

The temperature range for this mixing reaction has been observed by Corbett & Leach as being within the Epithermal zone. It is within this epithermal setting that the geothermal brines which contain the neutral chlorides mix. It has been stated before that *Sulfate salts and chloride salts are able to dissolve one another* which may have a positive affect on opal formation.

Salts are inhibitors of opal formation due to the fact that they act as catalysts to gel formation. When silica spheres gel before they grow then opal will not form. Opal spheres are roughly between 100nm and 800nm.

Salts that are in the solution when opal is forming must be neutralized or taken out of solution. This has been puzzling because the Australian opal environment is rich in salts. How does nature remove these salts from solution?

The simple answer to this is that certain organic chemicals precipitate salts. Ethanol is one of the most efficient, causing spontaneous precipitation of ionic salts. Most organic solvents will have the same effect. Acetone and Isopropanol should also work well. Organic solvents are freely available in the opal formation environment.

Silica Chemistry

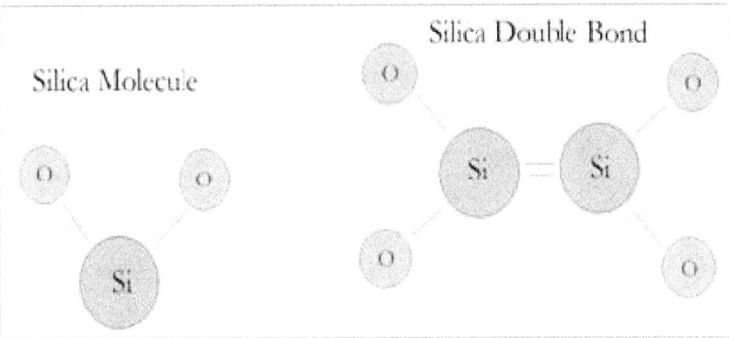

Figure 7. Shows the silica molecule & its double bond which makes it difficult to dissolve, much dissolved silica may not be single particles but two three or even clusters, creating different silica particle sizes. Salts are used to break this double bond to achieve single silica spheres rather than clusters.

2.48 Silica

Over the past 30 years there has been a growing interest in silica from both the scientific community and industry as a building block that may rival carbon in its versatility, however due to its great abundance it is expected to be cheaper in a large variety of applications.

The biggest hurdle in developing silica has been due largely to its high melting point, making it difficult to dissolve. Hydro Fluoric [HF] acid has been used to dissolve silica in the past but this acid is extremely dangerous to human life and health. Modern research has led to advances in the use of organic solvents to dissolve silica in much lower temperatures and is less toxic than HF acid.

Coupled with the use of organosilicon precursors such as TEOS and TMOS new products and applications have been developed. From this new research into silica and silicon have arisen entirely new boom industries such as:

- Nanotechnology
- Photonic Crystals [Inverse opals]

The Fibre Optics industry has also benefited from this research. You only need to do an internet search to appreciate the enormous extent of research that is currently being done on silica & silicon in a search to discover more products and applications for this element. The frontiers of mystery surrounding this element are being laid bare by the pioneers who are in the forefront of this research. The mysteries of Opal itself are no longer a mystery within certain scientific and research groups. Some universities in the USA are now even offering Opal Formation as part of their curriculum, possibly related to Photonic Crystals and Nanotechnology.

Silica is one of the most abundant minerals on the face of the earth which is of course the main ingredient in Opal. So you may then ask 'Why is opal not everywhere?

Silica [also known as silicon dioxide] is an oxide of the element Silicon.

Silicon is designated by the symbol Si in the table of elements. Oxygen has the symbol O_2; therefore the chemical symbol for silica is SiO_2.

However, Silica exists in a number of different forms which are:

- Dissolved silica H_4SiO_4,
- H_2SiO_4 Silicic Acid
- TEOS formula $Si(OC_2H_5)_4$

- Oligomers -- Polysilicic acid hydrated active silica spherical 50 A. These are described as Polymers.
- Colloidal silica = more highly polymerized species 50A plus.
- Silica Sol = 100nm or larger particle size, also known as either Polysilicic acid or colloidal silica.
- Quartz SiO_2 or Silicon Dioxide (Silica)
- Tridymite is crystal phase of Silica
- Christobalite is a crystal phase of Silica.

2.49 Silica facts:

- Silica does not dissolve easily.
- Opal formation needs an electrolyte
- Inorganic chemicals require either extreme acidic or high alkaline solutions to dissolve the silica.
- Evidence indicates that low chloride content is needed.
- Opal requires water to be chemically bound to the silica, which infers polymerization, which may require an initiator.
- Forms in clays
- Requires the presence of Alkali feldspar
- Pressure is required for drying, prevents cracking & crazing
- Organic chemicals are more effective in dissolving silica; some are able to do so at room temperature and at neutral pH.

These are some of the conditions under which opal forms. Now you can see why Opal does not form everywhere. If these conditions are not met then the chances of sedimentary opal forming is very small.

Table 4 Solubility of Amorphous Silica

Solubility of Amorphous Silica in water at 25°C at different pH	
pH	Solubility [ppm]
6-8	120
9	138
9.5	180
10	310 [Iler, 47]

The upper limit of temperature for amorphous silica is 250°C. As above this temperature the silica changes phase, and is no longer amorphous.

Precious opal does not form by simple silica deposition, or even by the deposition of dissolved silica. The formation of silica requires the chemical process for particle growth and for polymerization to bind the water molecules to the silica.

Opal is therefore a polymer i.e. the product of polymerization. Even when you look at opal, it looks like a polymer. Polymerization is proven out in the way that opal grows in seams, and by its chemically bound water. Opal forms in seams too often for it to be a coincidence. This is in fact the way that opal normally forms. The long held views that opal formed by deposition and simply filled cracks and voids where it hardened has been disproven by Len Cram. If it fills cracks and voids it is because the opal seams have grown into them.

Clues given by Len Cram:

- All starts with light (sunlight) [Plant? Organics]
- He adds TEOS, and electrolyte to the opal dirt
- It is an ionic polyelectrolyte
- There is an organic component.

2.50 Some Silica Sources

The reason for looking at silicification is because the dissolved silica that becomes opal must originate somewhere. Finding the source of the dissolved silica may well point the way to the method by which opal formation takes place.

Possible silica sources are:

- Water borne silica dissolved by hydrolysis, from silicate rocks, such as olivine $(MgSiO_4)$, may also contain Iron (Fe).
- Silicification; silica dissolved from mineral sources by the action of alkalis or by acids associated with hydrothermal events.
- Dissolved in situ by Organic reagents (Humic/Fulvic acids?). Amino acids, potassium feldspar?

- Volcanic Ash
- Biogenic, contained in all living things.

Most of the old theories concerning opal formation indicate that the opal is the result of hydrolysis with the silica being transported to the site where it settles over a period of time, this is referred to as sedimentation & eventually opal forms.

2.51 Defining Silicification:

This is the process that allows silica to dissolve and enter into other reactions. It probably involves an ion exchange process at the surface of the silica, removing metal ions that have shielded the silica making it virtually insoluble. Now exposing the silica surface to the full power of the alkali's, thus allowing the silica to dissolve.

Silicification can occur with a number of different electrolyte systems, but the electrolyte that we are interested in is the chloride electrolytes in the K-Al-Si system.

The following quote defines the role of Potassium feldspar in the silicification process.

> "The dark blue-gray material at left and below the vein is silica-carbonate rock (opalized serpentine). The gray material at right is adularized Knoxville formation mudstone. The term, "adularized," refers to the silicifying agent in the mudstone. In zones where the mudstone was hardened due to silicification, the responsible mineral was found to be the potassium feldspar, adularia.".(Ref: 36; D. Enderlin scan).

Adularia is the pure or nearly pure form of (orthoclase) potassium feldspar. It often has a pearly luster.
It is also called moonstone. Formula: $KAlSi_3O_8$

Clay reactions often happen in a hydrothermal environment and are similar to the alteration processes that occur within sediments. Different sources attribute these alteration processes to both hydrothermal and sedimentary environments. This is probably because the sedimentary layers often contain groundwater and are thus hydrothermal in nature.

Old theories of opal formation are now being challenged as they fail to account for the specifics of Opal formation. These theories state that opal is formed by soluble silica being deposited in crevices and holes within rock substrate where it settles and hardens over time. However this does not agree with the process of Polymerization, or ion exchange. Nor does it account for Opal growing in seams as proven by Len Cram.

The old theories also rely upon the right particle sizes collecting in the same place. This seems lame. It is more likely that the silica actually originates from the Feldspars and clays which are already weathered to a uniform size ready to enter into the silicification process.

"Figure 3.6 in the text (p. 45) shows the relationships among several important alteration types in the K-Al-Si system in a chloride electrolyte fluid. Reaction [1] marks the replacement of K-feldspar by muscovite and quartz while reaction [2] tracks the replacement of muscovite by kaolinite. In both these reactions hydrogen ions are consumed and potassium is released. The hydrolysis changes the pH and thus affects the stability of chloride complexes in the fluid. The released silica can either be immediately precipitated as silicification or migrate to form quartz veins in the vicinity".

(Wall rock Alteration and Gangue Mineralogy p45; ref: 35).

It is already agreed that opal forms in an aqueous environment. This aqueous environment is most likely to be a sedimentary in origin. The fact that these environments, hydrothermal, sedimentary and volcanic all share the sulfate chloride and carbonate salts indicates that these same salts may play an important role in silicification, the release of soluble silica often in bulk quantities, or at least has done so in the past. It is also noteworthy that certain salts when in aqueous solution have the ability to decompose one another by a process of double interchange of bases and acids. It seems in particular that sulfates affect the solubility of Chlorides. When carbonates of lime enter into solution, the silica is driven out [precipitation].

The Great Artesian Basin is the result of a sedimentary process. In view that this was once a great inland sea which has been built up by deposition of eroded weathering products and some uplifting, it can be assumed that the aqueous solution was salt brine, with similar salt content to sea water.

Temperatures in the Australian Opal Fields are often around the 40°C and in some places 50°C in the summer months. Temperatures for silicification of silica maybe slightly higher than this, but at this temperature we know that the carbonates would drop out of solution as insoluble complexes, this may allow the silica to dissolve more readily.

2.52 Alkali ions dissolve Silica

We know that Potassium is one of the major alkalis involved in the silicification process. In the natural environment Potassium Feldspar is often called Adularia, and the process of silicification is also referred to as adularization. This is because it is the K-feldspar that initiates the silicification process.

Groundwater also contains, other alkali ions, or acids that act as electrolytes allowing atom (ion) migration. This is the beginning of the ion exchange process which occurs particularly in the different clays and feldspars. See next section.

Elizabeth Smith claims that for silica to precipitate in nature from a solution that there are three major factors required:

- Alteration from alkaline to acidic environment
- The presence of oxides of Aluminium, iron or magnesium &
- The presence of NaCl or $NaSO_4$

Aluminium oxides are avail from the Kaolin, and the Finch Claystones which host opal in lightning ridge are slightly saline. The opal dirt in many parts of Lightning ridge contains calcareous material similar to limestones $CaCO_3$ locally known as white horse. Opal often occurs in association with carbonates.

2.53 Organic Silica [Biogenic Silica]

Recent research on Biogenic silica mainly related to diatoms and marine sponges has revealed that Polyamino acids also affect silicification. There is yet more research to be done in this area, more specifically in biosilicification, which relates to organic systems. Most living things contain Biogenic silica.

Table 5: Main Type of Silica Bonds	
Siloxane [Silicone]	**Silanol**
pH Acidic	pH Alkaline
Formula Si-O-Si	Si-OH-Si
Properties	
PolySiloxanes: This is the only type of Silica that is referred to as an ion exchanger and as an ion migration electrolyte. [See below]	

Silicones - Siloxane & Polysiloxanes [R_2SiO]

Siloxane bonds; Si-O-Si are linkages of Silicon and Oxygen and an organic molecule [R] of the Alkane family. PolySiloxanes contain more than one Alkane group. PolySiloxanes are ion migration electrolytes.

Alkanes belong to the hydrocarbon family [aromatic or aliphatic] associated with petroleum production. In fact Mercaptans [Thiols] are found in crude petroleum.

Some Alkanes are:

Paraffin Oil (English) = Kerosine (American).

Mineral oil, Vaseline, Petrolatum, Petroleum Jelly.

Phenyl or Phenol.

Methane CH4.

Thiols are produced by the decay of both vegetable and animal matter, the most common being Methyl Mercaptans CH3SH. Other Thiols are:

Coenzyme A.

Lipoamide.

Glutathione.

Cysteine [Amino acid].

Methyl Mercaptans.

Ethyl Mercaptans [Nucleophilic].

Sulfonic acid [RSO3H; See CH. 3] forms when Sulfuric acid and Alkanes mix. They are powerful ELECTROLYTES, used as ion exchangers. They also result by the reaction of Thiols with Sodium Hypochlorites [See: Hypochlorites CH. 2]. This could be the OPAL PETROLEUM LINK.

Aluminium Chemistry

2.54 Alumina

Aluminium is a soft metal with the chemical symbol Al. It is not known to exist as the pure metal in nature because it is very reactive and combines with other elements. My internet research has revealed that there exists some controversy over the specific nature of the form that Aluminium takes in clays; some think that it is Aluminium Oxide Al_2O_3 however others indicate that it is the monatomic form of the metal aluminium. This debate is however irrelevant for the purposes of this study, as it is known that Aluminium in solution exists only in one form that of Al_3+, the aluminium ion. An important property of the Al_3+ ion is that it hydrates vigorously in water solution. The formula for Aluminium in water can be written as $Al(H_2O)_6$ although it may also be written as $Al(OH)_3$ or Aluminium Hydroxide, when fully hydrated.

See: Figure 5: Alumina Hydration [Ref: 37]

Clays are Aluminosilicates, they contain mostly Aluminium and Silica. The bond formed is very strong. It is necessary to break this bond to form Opal. The silica must be released to form silica gel [opal gel], the Aluminium then being released of its bond to Silica will then rapidly hydrate to form Aluminium hydroxide [see reaction formula above] which is a polymerizing agent for silica.

Both the Al(OH) 3 [Polymerizing agent] and the H+ ion are essential for the proper formation of opal. H3O+ is normally written and expressed as H+. The function of the H+ ion is in the bringing together of the silica in a process called flocculation. The H+ ion is also produced by Carbonic Acid and other acids. It is the H+ ion that is most effective in releasing Aluminium from clays in solution initially forming the Al3+ ion which rapidly hydrates to Aluminium Hydroxide [Al(OH) 3].

Water in an acidic solution will have more H+ ions than OH- ions but in alkaline solutions water contains much more OH- than H+ ions. This is a very important factor because H+ ions dissolve Aluminium and OH- ions dissolve silica.

2.55 Breaking the bonds
How then do we break the Si-Al bonds?
One way to break the bonds is to dissolve one or both of the elements:

- Both HCL → Aluminium Chloride
- Sulfuric acids → Aluminium sulfate.
- Salt [NaCl] ; salt water corrodes it rapidly esp. when HOT
- Carbonate Reactions [See Ch. 2 & 4]

Aluminium also reacts with strong Alkali, such as Sodium Hydroxide which forms Sodium Aluminate and releases hydrogen gas:

$$2Al + 6 NaOH \rightarrow Na_3AlO_3 + 3H_2$$

A little Sodium Hydroxide added to an Aluminium salt can

form Aluminium hydroxide:

$$NaOH + AlSO_4 \rightarrow Al(OH)_3 + NaSO_4$$

Also Soluble Carbonates (see below) or Soluble Sulfides added to an aluminium salt form Aluminium Hydroxide.

Aluminium may also be dissolved by CO_2 or by H_2CO_3 Carbonic Acid. This does not combine with the Al_3^+ but releases it into the water solution where it hydrates.

Aluminium Hydroxide tends to form gels, often brown in colour in solution.

The significance of Aluminium & Oxides/hydroxide will become better understood by reading:

- Ch. 2 Aluminium Chemistry:
- Ch. 2 & 4 Carbonate reactions [Carbonic acid & dissolved silica]
- Ch 4 Polymerization

2.56 Aluminium Sulfate Reaction

[Ref: 38; pg 279]

In solution Aluminium Sulfate $AlSO_4$ tends to be acidic due to hydrolysis. If reacted with Sodium Hydrogen Carbonate it releases carbon dioxide but the intermediary product is Carbonic Acid.

"$2Al^{+++} + 3SO_4 = 4H_2O \leftrightarrow 2Al(OH)^{++} + 2H_3O^+ + 3SO_4 =$

$Na^+ + HCO_3^- + H_3O^+ \leftrightarrow Na^+ + H_2CO_3 + H_2O$

$H_2CO_3 \rightarrow H_2O + CO_2 \uparrow$".

[Quote Ref: 38; pg 279, Modern Science by Metcalf, Williams and Castka 1966]

2.57 Aluminium as Dehydrating Agent

"Activated alumina, Al2O3 made by moderately heating aluminium hydroxide, is an effective adsorbent for water vapor. It is useful in removing water vapor from various gases. Activated alumina is often used in the chemical laboratory as a desiccant (Ref: 38, H. Clark Metcalf; John E., Williams; Joseph F.,Castka, page 255)".

2.58 Role of Alumina

As the opal gel dehydrates the $Al(OH)_3$ also begins to dehydrate along with it, resulting in Al_2O_3 forming which further dehydrates the opal gel. Sulfuric acid is often the agent which forms carbonic acid from carbonates. When Calcium and Sodium carbonates occur together in the opal environment and are acted upon by sulfuric acid, they form carbonic acid and the salts Calcium sulfate and sodium sulfate which can combine under the influence of the acid to form Glauberite. The glauberite itself acts like a desiccant soaking up water. This helps in the process of dehydration, which must occur for opal to form.

However I do not agree with Elizabeth Smith that it is essential for opal formation. I can see how it could assist the process of dehydration but there are other processes at work that will produce dehydration. Glauberite is rather evidence of Sulfuric acid alteration. The hot humid conditions of the opal environment also make a large contribution to the dehydration process.

2.59 Alternative Aluminium Hydroxide Production

Sodium Hydroxide plus Aluminium salts result in a [white] gelatinous precipitate of Aluminium Hydroxide [Al(OH) 3]

[Ref: 38; 471].

TABLE 6 – Dissolves or Etches Silica	
Phosphoric Acid H_3PO_4	Dissolves white quartz [SiO_2]
Polyamines [$x.NH_2$]	Putricine & Cadaverine
Hydrofluoric Acid HF	Dissolves silica [SiO2]
Potassium Hydroxide KOH	Dissolves Silicon & Al
Ferric Chloride FeCl	Used as Etchant in industry
Ammonium Chloride	Possible
Carbonic Acid H_2CO_3	RAPIDLY dissolves Silica
EDA & EDTA	Possible similar to Polyamines
EDP ethylenediamine Pyrocatecol &H_2O	Dissolves Silicon but not oxide BUT Poisonous
Carboxylic Acids [x.COOH]	Probably because they degrade to Carbonic Acid & CO_2 • Acetic Acid H_3COOH • Citric Acid • Tartaric Acid • Fulvic Acid

TABLE 7 – Dissolves or Etches Aluminium	
Phosphoric Acid H3PO4	Dissolves Aluminium enhanced by addition of Acetic acid [Ethanol degrades to ACETIC]
Hydrochloric Acid HCL	Dissolves Aluminium
Potassium Hydroxide KOH	Dissolves Aluminium
Sodium Hydroxide NaOH	Dissolves Aluminium evolves hydrogen gas
Ferric Chloride FeCl	Dissolves Aluminium may work better with a carboxylic acid such as Citric or Acetic or Oxalic acids
Ammonium Chloride	May also work with Carboxylic acids

Silcrete and Mound Springs

2.60 Silcrete forms concurrently with Opal

Silcrete is another clue to the process. Silcrete is almost always associated with Opal formation. Often called Duricrusts. It is noteworthy that Iron species also form duricrusts. The Iron species that maybe involved in the polymerization only as an initiator/catalyst and does not enter into the reaction, thus it remains as a byproduct, possibly becoming part of the silcrete.

Silcrete (known also as "Pseudoquartzite" or gannister). This is a finely crystalline rock composed of silica that commonly contains fossil stem casts. An erosion-resistant material that occurs mainly as a caprock on buttes or as scattered bounders. True quartzite is a metamorphic rock; since this rock was probably deposited in water as a silica gel, it is referred to as pseudo-(like a) quartzite. Age can range from late Cretaceous through Tertiary. (ref: 49; North Dakota Geological Survey)

Following is a description of gannister:

GANISTER OR GANNISTER. — A very hard and siliceous fire-clay found beneath some of the seams of the lower coal measures. It contains stigmaria roots. It is manufactured into fire-bricks of a highly refractory character, which can be imitated by grinding and mixing up fire-stone post with a less refractory clay. This post is also called ganister. (ref: 50; The Durham Mining Museum).

The same thing happens with Len Cram's opal, a cap forms on

it. This is proof that it is one and the same process. Opal formations are almost always capped with silcrete. In some instances the silcrete has been moved from its position probably by earth movements over a long period of time but is normally found not far from an opal seam. The caps of mesas are often silcrete. Finding silcrete is a very good indication that opal is near.

Opal deposition and silcrete formation were probably contemporaneous. (ref: 48; Opal Hut).

Even though I believe that Acid Sulfates are linked to the opal forming electrolyte, I would not rule out the possibility of some chloride involvement in the process as chloride sulfate brines are common in opal areas. Yet the NaCl may only be linked to potch formation by deforming the silica spheres. The role of NaCl remains as yet uncertain; however it is known that precious opal and potch often form together.

2.61 Analysis of Silcrete

The formation of silcrete is still under debate, as are origins of the silica, its movement, and the environment or chemical processes causing the later alteration, deposition, or precipitation. Chemical analysis has been done on a wide variety of silcrete, and from a large range of environments.

Silcrete typically contains:

- silica to 85% and:
- aluminum, Al_2O_3 3% approx.

- iron, Fe_2O_3 2% approx.
- *titanium* TiO_2 up to 10%[Anatase]

TiO_2 may be contained in significant quantities, with more of titanium, than aluminium or iron.

The climate under which it forms remains uncertain, as almost all silcrete found are "fossil"--they formed in the past, with little forming today.

The analysis above gives a rough estimation of the oxide ratios that may be catalysts in opal formation. Keep in mind that these are catalysts and would represent only a minor portion of the opal ingredients.

2.62 Silcrete ingredients

- Pigments all the following are used as paint pigments and stains:
  - Fe_2O_3,
  - TiO_2 along with
  - Limonite (yellow) same formula goethite (hydroxide).

2.63 Two types of Silica

In opal formation there are two types of Silica involved. One type of silica is polymerized as opal, the other is rejected and along with the photo catalysts becomes part of the Silcrete, acting as a binding agent.

The question then is why is one type of silica accepted and the other rejected. The Author believes the silica that goes into the opal formation is amorphous, very fine silica particles, and the rejected silica is larger size although fine size quartz crystals. Both types are crystalline but different sizes.

2.64 From Silcrete to Photo Fenton Reaction

The jump from silcrete to Photo Fenton reaction is commonsense when you know that nearly every type of Opal is found in an environment containing a lot of Iron. The clues to this reaction being involved in opal formation are:

- Silcrete Photocatalysts (Photo Fenton)
- Opals Iron environment (Iron Species - Fenton reaction)
- Sulfuric Acid (Fenton catalyst)

It is also possible that the silica portion of the silcrete may also be a photo catalyst (unconfirmed). Some silica types are used as photo catalysts.

2.65 Photochemical Reactions (Solar/Light reactions)

Some photochemical reactions are now being called Advanced Oxidation Process (AOPs).

It is also very significant that most of the silcrete materials fit into a category called photo catalysts. Most metal oxides can be used as photo catalysts.

> *"Photocatalysis can be defined as the acceleration of a photoreaction by the presence of a catalyst"* (Ref: 52; Stephen Popielarski).

TABLE 8: Photocatalysts [See Silcrete]		
Main Photocatalysts	**Activators**	**Other Photocatalysts**
Fe_2O_3,	Visible Light (VIS)	CuO
TiO_2 Anatase	Ultraviolet Light (UV)	ZnO
Al_2O_3	Iron II i.e. Transition metals	Manganese Dioxide MnO
SiO_2	H_2O_2	SO_4 (Sulfates)
Sodium Thiosulfate		

Photocatalysts in solution are activated by sunlight or UV light. UV light is probably more efficient. Len Cram mentioned that it all starts with light. By far the most widely used photo catalyst is TiO_2.

2.66 Mound Springs [Ref: 54]

Byron Deveson, geochemist and a Director of Opal HiTech whilst studying mound springs discovered a link between mound springs and silcrete. Most, that is to say nearly all mound springs have boulders of silcrete near the edge or boundary of the springs. He describes how alkaline mound springs waters arising from deep fractures in the earth, typically running along fault lines and linear splays are involved in the genesis of opal. His findings were published in 2002 under the title "The Origins of Precious Opal – A New Model" in "The Gemnologist" Vol 2, 2002.

Mound Springs are also known are 'natural artesian springs', however the term mound springs is much more prevalent in literature and scientific circles.

Opal tends to be deposited close to fault lines and blows which are believed to be pathways for water saturated in silica to flow along.

Mound Springs are a major natural environmental feature throughout the opal fields of Australia and for this reason alone must be studied in relation to opal formation in the Australian opal fields.

The Lightning Ridge opal fields can be plotted on a map showing that they are in-line with a string of springs that run from as far north as Dirranbandi in Queensland to as far south as Coolibah springs in NSW [300kms].

Another group of mound springs that lies outside the boundaries of the GAB called the Corea [Koree] group of mound springs located about 30kms north of Nymagee. This group of springs runs for approx 100kms.

Many of the mound springs of the GAB area are still active whilst others are now extinct. The inactive mound springs have simply dried up over time but still remain a feature of the landscape, recognized by the rounded shape and the silcrete boulders scattered around the boundaries.

The Lightning Ridge opal fields and mound springs run along a line 20 degrees magnetic, parallel to underground features such as fault lines.

"A plot of sedimentary opal fields shows that NSW and South Australia they are located near or on the intersection of continental scale lineaments, or a short distance along significant splays from these linear features." [Byron Deveson, pg 2]

2.67 Formation of Silica Spheres in Mound Springs

On the opal fields within the sediment water patches are known to have concentrations of dissolved silica ranging from 100-120 ppm. In hot springs however, the concentrations are typically 300-500 ppm and have been recorded as high as 700 ppm.

It is common knowledge that precious opal is an array of uniform amorphous silica spheres in a cubic arrangement in the size range of 0.14 to 0.30 microns in diameter.

Such spheres have been prepared from geothermal waters [alkaline waters of pH 8.7, nearly saturated with silica]. The process involved firstly cooling, then lowering pH, followed by ultra filtration and dialysis.

This experiment was conducted in 2000 at Wairakei power station by Brown and Bacon in New Zealand.

The above experiment provides evidence that precious opal could be formed by mound spring waters in a very similar fashion. These mound springs contain the right chemical mix and pH for silica hydrosols to form and then be concentrated by ultra filtration and removal of impurities by clays, this process will be explained later as it is essential to natural opal formation. [Montmorillonite].

> *"Siliceous sinters formed in terrestrial hot springs under high temperature (75-100C), reducing, and silica saturated conditions and are thought to be formed in conditions similar to that of early Earth (Konhauser et al., 2003; Lynne et al., 2006). Sinters are of interest because they contain active sites of microbial activity and biomineralization. The transition from disordered opal-A ($SiO_2.nH_2O$) to more ordered forms (e.g. opal-C, opal-T, opal-CT, quartz and morganite) can affect the preservation potential of microbes such as hyperthermophilic Aquifex and Thermotoga, which are known to live in modern environments (Blank et al., 2002)."* [Kyle & Schroeder, 189].

2.68 Factors affecting Sphere Size

Several factors affect both the uniformity and the ultimate size of the Opal spheres. These factors include the following:

- The starting and finishing pH.
- The initial concentration of Silica
- Concentration of any silica complexing agents.
- The starting and finishing temperatures.
- The presence of any hydrosol stabilizing agents, includes surfactants.
- Ionic strength.
- The initial concentration of nuclei in the mix.

- The charge of the nuclei. Silica which is negatively charged will be attracted to positively charged nuclei.

Precious opal contains many trace elements including nucleating agents which include:

- Zirconium
- Zirconium Hydroxide
- Zirconium Phosphate, absorbs cerium, thorium and Titanium.
- Aluminium
- Titanium
- Thorium

2.69 Changes in Mound Spring properties

Suitable changes in physio-chemical properties of any mound spring with the appropriate water composition [almost saturated with silica at high pH] could cause silica spheres to form. The expected changes in a mound spring environment are normally; pH; ionic strength; and temperature of solution.

> *"Mixing warm, alkaline mound spring water with cool, slightly acidic ground water with low total dissolved salt content would, decrease pH, lower the temperature, and lower the ionic strength of the mound spring water. All three changes favour the formation of Silica spheres."* [Byron Deveson, pg 5].

2.70 The Role of Clay

Opal dirt is found throughout the Australian Opal fields, considered by some [Elizabeth Smith, 48] as simply a mixture of clays, yet by others as a clay stone containing considerable amounts of swelling clays. In Lightning Ridge the swelling clays are mainly of the Montmorillonite type of clay.

Montmorillonite clay is thought to favour the formation of silica spheres and also has special properties which help in the opal genesis process. The clay due to its superior ion exchange capacity is able to perform the functions of ultra filtration and dialysis of the water which carries the silica spheres.

Montmorillonite can:

- Act as a semipermeable membrane [often lining cracks].
- As an ion exchanger, it is a strong cation exchanger which absorbs impurities from water, producing a pure source of water.
- Exchanges impurities for ion such as Sodium and Potassium.
- The proton form is able to render the mound alkaline spring water neutral. Alkali neutralizing capacity.

Other clays can function as ion exchangers or as semipermeable membranes but Montmorillonite is superior to them all in these functions. Semipermeable membranes have the ability to concentrate and purify dilute solutions of suspended silica spheres.

Another very important function of montmorillonite and other swelling minerals is their ability to pressurize fluids containing silica spheres. This pressurization works either by the absorption of water along with silica sphere growth or by the exchange of cations.

It is extremely probable that this pressurization is the motive force for the ultra filtration of the fluids. The force is sometimes so great that it causes an explosion rupturing the rock walls, sometimes forming blows. The pressure produced by this mechanism can be as high as 50 psi [3.45 bar] which is generated in open systems due to water absorption by the montmorillonite. It has been known to happen around building foundations.

Clays are ever present in the sediment structures of the opal environment. Bulk clay, however is not a requirement to act as a semi-permeable membrane. Even a thin layer of clay, or silica gel, lining a crack or crevice can act as a semi-permeable membrane.

Montmorillonite also performs a number of other functions. Precious opal also referred to as Opal A in the presence of the Mg ion or carbonates changes phase to Opal CT, which does not form precious opal.

Montmorillonite can prevent this phase change by absorbing the Mg or carbonate ions, thus increasing the chance of opal forming.

Microbes also find montmorillonite to be a favorable habitat allowing increased rates of growth.

2.71 Trace Elements of Opal

As mentioned before opal can contain a myriad of trace elements. These trace elements must have an originating source. Most of the opal models put forward so far have not

been able to accurately account for these trace elements found
in opal.

There are a number of elements present in opal in quantities
that are significant despite being unreactive, some of these are:

- Zirconium
- Hafnium
- Niobium
- Tantalum
- Thorium
- Titanium

The source of these trace elements could not be accounted for
if the silica source was solely weathering fluids. Mound spring
waters however, contain intrusive material such as Carbonatites
and Natrocarbonatites which gives a credible account for the
trace element observed in opal.

The presence or absence of various trace elements in precious
opal is due to alkaline hydrothermal leaching adding trace
elements to the source waters, followed by processes such as;
pH neutralization, and dialysis which removes most impurities.

2.72 Clays as Semi-permeable Membranes

Forcing materials through semi-permeable membranes requires
some type of pressurization. Deveson has proposed that the
pressure may be caused by a mechanism of hydraulic pumping
caused by seismic or other ground activity such as geyser
activity, or mud volcano activity pumping pockets of silica sol
material into a system of voids and cracks at intermittent
intervals. These voids were caused by previous leaching of
softer rock, fossil or evaporite materials in the sedimentary
layers or perhaps by hydraulic fracturing.

This is thought to operate at shallow depths, possibly less than 100 metres due to the need for open spaces within the sediment.

The other possible way that a suspension of silica spheres could be pressurized is by the swelling of Smectite clays when they absorb water or when they absorb ions such as sodium. Read describes the Mound springs in the Eulo area as oozing blue green mud [Celadonite?], like some mud volcanoes where natural gas is the driving force.

2.73 Purification of Silica Spheres by Dialysis

The driving force for dialysis of suspended silica spheres would require either the input of a counter current of water of lower ionic strength. Simple pressurization is sufficient and can easily be generated by a mound spring. Intermittent violent eruptions in some mound springs is all the evidence needed to show that mound springs can become highly pressurized.

2.74 Exploding Mound Springs

An outstanding feature of Lightning Ridge opal is its unusual association with breccia pipes, also known as blows. Breccias are not just common to Lightning Ridge but are also seen on the other Australian opal fields.

Artesian springs in other parts of the world can also form breccia pipes, yet it is not known how they form. Mound springs in the Eulo-Yowah district and also the mound springs near Malpas Queensland have been known to occasionally explode ejecting gravel and boulders up to 600mm in diameter.

2.75 Existing Models and Compatibility

There have been three main models that have been accepted for sedimentary opal formation. The models are: the microbial; the weathering model and the syntectonic models.

The model presented by Byron Deveson actually includes all three into a single model, but with greater emphasis on extending the Syntectonic model. Deveson has provided evidence that mound spring activity and the breccia pipes of lightning Ridge were much more recent than originally thought, by carbon-14 dating showing that samples were approx 4000 years old which is in agreement with the Syntectonic model by Pecover [2003].

2.76 Bacterial Opal Link

K. Dowell and Mavogenes reported fossil bacteria in opal samples. Clay stone, which hosts opal commonly contains high amounts of fossilized organic material that provide nutrients for bacteria. The probability that bacteria inhabited mound spring water, even semi-stagnant mound spring water is very high. Mound springs provide large amounts of nutrients that could feed large colonies of bacteria. Nutrients, such as:

- Ammonia
- Nitrates
- Phosphates

Sufficient nutrients exist, leaching from the GAB to cause bacteria to flourish, in this nutrient rich environment.

Bacteria are known to excrete organic compounds. These compounds have surface active properties. These surface active compounds act to enhance silica growth and stabilizing silica spheres. Abnormal growth of silica scale on the Wairakei geothermal power station has been blamed on the presence of bacteria. This is not unusual as silica gel has been a culture medium for bacterial growth experiments for many years.

One of the main chemicals excreted by most bacteria is Ammonia NH_3 which is hydrolysed to the ammonium ion NH_4. This chemical is believed to be one of the main catalysts for opal formation.

2.77 Boulder Opal and Volcanic Precious Opal

Active mound springs appear near some of the boulder opal sites [Lila springs]. Breccia pipes do not occur as a common feature of boulder opal therefore the filtration must have been driven by sustained pressure from mound spring waters, rather than the rapid high pressure events that seem to be common at Lightning Ridge.

Silica Spheres, and then linear chains of these spheres, would attach themselves to ironstone surfaces and within voids in ironstone boulders. This occurs as hydrated iron oxides are normally oppositely charged to the silica spheres.

Deveson indicates his theory may not be able to explain volcanic opal formation. Despite the fact that waning volcanic systems emit hydrothermal fluids, it is generally thought that volcanic hydrothermal fluids contain too much material that would prevent or inhibit the formation of opal.

On the other hand it is known that when Acid sulfate hydrothermal fluids from volcanoes mix with cooler neutral chlorides waters that opal will result.

Figure:8 Role of pH

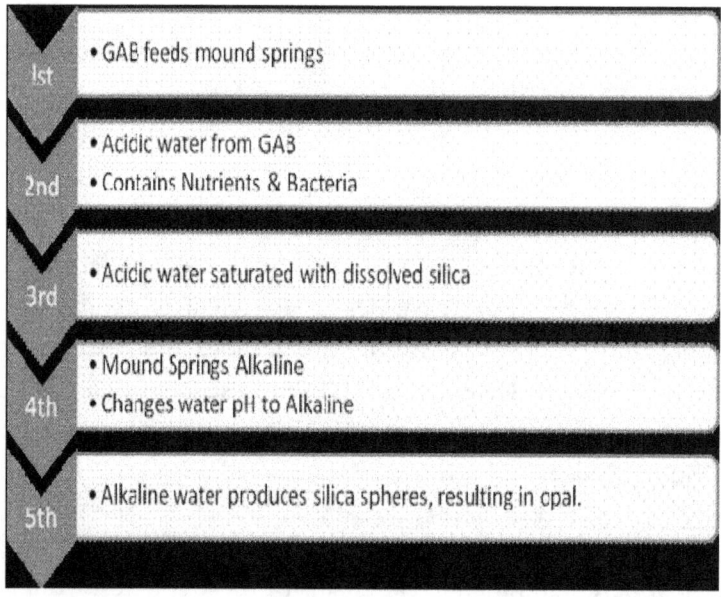

Related Topic: Weathering

Clays; hydrous oxides; aluminosilicates; geology; soils; regolith; laterite; weathering; rocks & minerals; sedimentary deposits; dinosaurs; petrology; alteration; hydrothermal alteration; argillic; Silicic; sericitic; polymorphology; authigenic; propylitic; hyaloclastite; volcanic breccia; Polytype; Phyllosilicates; Detrital Minerals; glauconite; celadonite; ion-exchangers; anion; cation; metal complexers; hydroxyl.

Related Topic: Clays

Erosion; Earth Sciences; Rocks & Minerals; Soils; dissolution; hydrolysis; oxidation; reduction; deposition; sediment; laterites; landslides; mudslides; regolith

Related Topics – Sulfates

Oxidation; reduction; anions; cations; ion exchange; base exchange; electrolytes; hydrogen; electron donors; electron acceptors; reagents; reactants; oxidizers; speleogenesis; alteration; hydrothermal fluids; hydrothermal electrolytes; acid sulfates (Bisulfates); etchants, geoscience, hydrolysis (reaction of salts).

Related Topic - Silica

Silicic alteration; silica dissolution; dissolved silica; silica reactions; silicates; aluminosilicates; hydrous oxides; petrification; opalization;

Polymorphology; silicified wood; acid sulfates; bisulfates (hydrogen sulfates -- acid sulfates); thiosulfate; Alkali-Silicate reaction.

[For information on Carbonic acid and Phosphate reactions and how these reactions dissolve Aluminium; see CH. 3.]

Related Websites

Hydrous Oxides

http://grunwald.ifas.ufl.edu/Nat_resources/silicates/silicates.htm

Other Resources

Ask a scientist websites

- US Geological Survey Information
  http://geology.usgs.gov/index.shtml

- Ask an Earth Scientist from the Dept. Geology & Physics at the University of Hawaii
  Http://www.soest.hawii.edu/GG/ASK/askanerd.html

Further Reading

Cuadros, J., Altaner, S.P. (1998) Characterization of mixed-layer illite-Smectite from bentonites using microscopic, chemical, and X-ray methods: Constraints on the Smectite-to-illite transformation mechanism: <u>American Mineralogist</u>, 83, 762-774.

Altaner, S.P., Ylagan, R.F. (1997) Comparison of structural models of mixed-layer illite/Smectite and reaction mechanisms of Smectite Illitization: <u>Clays & Clay Minerals</u>, 45, 517-533.

Ylagan, R.F., Altaner, S.P., Pozzuoli, A. (2000) Reaction mechanisms of Smectite Illitization associated with hydrothermal alteration from Ponza island, Italy: <u>Clays & Clay Minerals</u>, 48, in press.

The Smectite to chlorite transition in the Chipilaga geothermal system, El Salvador, by D. Robinson and A. Santana DeZamora.

Clay Mineralogy [Library Ref: 552.52]

Ralph E., Grim

McGraw-Hill, New York, 1953

Data handbook for clay materials and other non-metallic minerals

Pergamon Press, Oxford; New York, 1979 [Library Ref: 549.6 D232]

Applied Clay Mineralogy [Library Ref: S620.191]

Ralph E., Grim

McGraw-Hill, New York, 1962

Chemistry of Clay & Clay Minerals [549.67 C517]

Ed., A.C.D. Newman

Harlow; Longman Scientific & Technical 1987

Bibliography:Chapter 2

1 *Link to Rocks & Minerals*

Http://www.yahooligans.com/science_and_Oddities/Earth_T he/Geology/Rocks_and_Minerals/

2 Title: *Unit 2 - Geochemical methods*

Subtitle: Clay Reactions - geochemical sampling Introduction Resources

Subject: Alteration Mapping

Delta Mine Training Centre Alaska

Http://www.dmtalaska.org/exploration/ISU/unit2/u2lesson2 .html

3. *GEOL 243.3 Sedimentology: Course Notes (week 2) ctd.*

Subject: Sedimentology - Hydrolysis reactions of Silicate Minerals i.e. Clay reactions.

http://www.usask.ca/geology/classes/geol243/243notes/243 week2b.html

4 *Black Opals of Lightning Ridge*

Elizabeth Smith

Simon & Schuster; Roseville NSW 2069 Australia; 1999

5 *Clay Types* [2001 – 2003]

California Earth Minerals Corp.

http://www.calearthminerals.com/claytypes.htm

6 *The Chemistry of Silica, (Solubility, polymerization, colloid….*

Ralph K., Iler

Wiley, New York, 1979   [Pg 399]

7 *Sodic Soils and their Environmental Management*

Website Article
http://www.fao.org/documents/show_cdr.asp?url_file=/docr
ep/x5871e/

x5871e05.htm

8. *Feldspar Dissolution Kinetics in Organic Electrolyte Solutions --*
http://www.geo.utexas.edu/chemhydro/dissolution_kinetics/I
ntro.html -- Wan-Joo Choi at wanjoo@mail.utexas.edu

Saturday, May 02, 1998

Soil Reactions:

9. *Journal of Cave and Karst Studies* - ISSN 0146-9517
Volume 63 Number 1: 23-32 - April 2001
A publication of the National Speleological Society

By-product materials related to H2S-H2SO4-influenced
speleogenesis of Carlsbad, Lechuguilla, and other caves of the
Guadalupe Mountains, New Mexico
Victor J. Polyak and Paula Provencio

10. *Sulfur*

http://www.bae.umn.edu/extens/manure/programs/sulfer.ht
ml

Last updated July 1, 1998 by David Schmidt

The University of Minnesota

11. *Alkaline Silicate Reaction* (ASR)- 10.104

Advanced Cement Technologies
Technical Bulletin

Web source:

http://www.metakaolin.com/member/10.104%20Alkali%20Si
lica%20Reaction.htm

12. *An Introduction to Acid Sulfate Soils*

By Jesmond Sammut & Rebecca Lines-Kelly

University of New England, Armidale 2003.

13.   *Preliminary Geologic Map And Alteration Mineralogy Of The Main Scarp Of The Slumgullion Landslide Chapter 3*

by Sharon F. Diehl and Robert L. Schuster

Source: http://pubs.usgs.gov/bul/b2130/Chapter3.html

14.   *Mineral formation and redox-sensitive trace elements in a near-surface hydrothermal alteration system*

Source:
http://www.osti.gov/energycitations/product.biblio.jsp?osti_id=687712

Authors: Gehring, A.U.[ETH Zurich, Schlieren (Switzerland). Inst. for Terrestrial Ecology] | [ETH Zentrum, Zurich (Switzerland). Office of Planning]; Schosseler, P.M.; Weidler, P.G.[ETH Zurich (Switzerland). Inst. for Physical Chemistry]

15.   *Kaolins In North Sudan* [Abstract by Dr. Mario Wipki]

Source: http://mindepos.bg.tu-berlin.de/lager/wipki/abskurz.htm

Subject: Alteration of Kaolin to Alunite

16. *Sedimentary Rock-hosted Opal;*

in Selected British Columbia Mineral Deposit Profiles, Volume 3, Industrial Minerals, (1999)

Authors: Paradis, S., Townsend, J. and Simandl, G.J.

Editors, G.J. Simandl, Z.D. Hora and D.V. Lefebure,

Publisher: British Columbia Ministry of Energy and Mines
    by: S. Paradis[1], J. Townsend[2] and G J. Simandl[3]

> [1] Geological Survey of Canada, Pacific Geoscience Centre, Sidney, British Columbia, Canada
> [2] South Australia Department of Mines and Energy
> [3] B.C. Geological Survey, Victoria, British Columbia, Canada

17. *Precious Opal In Volcanic Sequences Q11*
by S. Paradis[1], G.J. Simandl[2] and A. Sabina[3]

[1] Geological Survey of Canada, Pacific Geoscience Centre, Sidney, B.C., Canada
[2] British Columbia Geological Survey, Victoria, B.C., Canada
[3] Geological Survey of Canada, Ottawa, Ontario, Canada

Paradis, S., Simandl, G.J. and Sabina, A. (1999): Opal Deposits in Volcanic Sequences; in Selected British Columbia Mineral Deposit Profiles, Volume 3, Industrial Minerals, G.J. Simandl, Z.D. Hora and D.V. Lefebure, Editors, British Columbia Ministry of Energy and Mines

18. *Hot-Spring Au-Ag H03*
by A. Panteleyev
British Columbia Geological Survey

Panteleyev, A.(1996): Hot-spring Au-Ag, in Selected British Columbia Mineral Deposit Profiles, Volume 2 - Metallic Deposits, Lefebure, D.V. and Hõy, T, Editors, British Columbia Ministry of Employment and Investment, Open File 1996-13, pages 33-36

19. Hot Spring Hg H02
by A. Panteleyev
British Columbia Geological Survey

Panteleyev, A.(1996): Hot-spring Hg, in Selected British Columbia Mineral Deposit Profiles, Volume 2 - Metallic Deposits, Lefebure, D.V. and Hõy, T, Editors, British Columbia Ministry of Employment and Investment, Open File 1996-13, pages 31-32.

20. *Epithermal Au-Ag-Cu: High Sulphidation H04*
by A. Panteleyev
British Columbia Geological Survey

Panteleyev, A.(1996): Epithermal Au-Ag-Cu: High Sulphidation, in Selected British Columbia Mineral Deposit Profiles, Volume 2 - Metallic Deposits, Lefebure, D.V. and Höy, T, Editors, British Columbia Ministry of Employment and Investment, Open File 1996-13, pages 37-39.

21. *Regolith 2004*

Opalization of Fossil Bone and Wood: Clues to the Formation of Precious Opal [Pages 264 – 268]

Benjamath Pewkliang [1,2], Allan Pring[2] & Joel Brugger[1,2]

[1]CRC LEME, School of Earth and Environmental Science, University of Adelaide, SA 5005

[2]Department of Mineralogy, South Australian Museum, North Terrace, Adelaide 5000, South Australia

22. *The Chemistry of Silica, (Solubility, polymerization, colloid and...*

Ralph K., Iler

Wiley, New York, 1979

23. *The Missing Organic Molecules on Mars*

Subjects: Kerogen (Derived from Coal)

Amino Acids & hydroxyacids (Ruff Deg...).

Http://www.pubmedcentral.nih.gov/articlerender.fcgi?....

24. *Sodic Soils And Their Management*;
http://www.fao.org/documents/show_cdr.asp?url_file=/docr
ep/x5871e/x5871e05.htm

25. *Indicators*

Sambal's Science Web

http://sambal.co.uk/indicators.html

26. *Feldspar Dissolution Kinetics In Organic Electrolyte Solutions* --
http://www.geo.utexas.edu/chemhydro/dissolution_kinetics/I
ntro.html -- Wan-Joo Choi at wanjoo@mail.utexas.edu

Saturday, May 02, 1998

Carbonates:

27. *Making Carbonic Acid*

Dr. Mabel Rodrigues

Article (Paragraph)

28. *High Temperature Water*

Shawn Hunter – The Savage Group

29 *Blue Bottle*

Website:
http://www.swintonmath.com/port/inst/blue_bottle.htm

30. *Speleogenesis*

(Section on Carbonic Acid dissolution).

www.speleogenesis.info/archive/sg2/Klimchouk2/index_print
.htm

Salts & Brines
31. *The New Penguin, Dictionary of Geology*

Second Edition 1996
Ed., Philip Kearey

New York, New York.

32. *Wikipedia, The free Encyclopedia*

Subjects: Felsic [relates to volcanic rocks]
Date research: 13/5/08

http://en.wikipedia.org/wiki/Felsic

33. *Hydrolysis: Acidic, Basic and Neutral Salts*

Page 1
Date research: 13/5/08
http://www.science.uwaterloo.ca/~cchieh/cact/c123/salts.ht
ml

33a. *Ionic Salt Precipitation* [Paper – 2 pages].

NEWTON Ask A Scientist [Vince Calder & Jim Swenson]
Chemical Archive 19/12/2005

Http://www.newton.dep.anl.gov/askasci/chem03/chem03689
.htm

33b. *Hyper filtration-Induced Precipitation Of Sodium Chloride*

By: T. M. Whitworth, Principal Investigator, NMBMMR, New
Mexico Tech

And: Chen Gu, Research Assistant, Department of
Environmental Engineering, New Mexico Tech.

February, 2001

Tr314.pdf   [page 4]

33c. *Salinity*

Wikipedia

http://en.wikipedia.org/wiki/Salinity

24/09/2007.

Silica & Aluminium:

34. *Kaolins In North Sudan* [Abstract by Dr. Mario Wipki]

Source: http://mindepos.bg.tu-berlin.de/lager/wipki/abskurz.htm

Subject: Alteration of Kaolin to Alunite

35. *Wall rock Alteration and Gangue Mineralogy*

http://www.geology.wisc.edu/~pbrown/g515/alteration.htm

36. Website Title: *Homestake Mining Company - McLaughlin Mine Mclaughlin Gold*

Subject: Silicification

Author: D. Enderlin scan

http://wwwnrs.ucdavis.edu/mclaughlin/naturalhis/gold/gold1.htm

37. *Weathering of rocks*

http://fbe.uwe.ac.uk/public/geocal/soilmech/classification/soilclas.htm

#CLASSWEATHER

38. *Modern Chemistry* [Pgs 255; 279; 469-471; 473-475]

Authors: H. Clark Metcalf; John E. Williams and Joseph F. Castka

Publisher: Holt, Reinehart and Winston, Inc. NY, Toronto, London

Copyright: 1966

39. *Virgin Valley Opal*

Subject: Geology -- Volcanic Ash -- opal
http://goldnuggetwebs.com/VVOPALS/GEOLOGY.HTM

40. *Origin of Volcanic Agate*

Dr. Jens Gotze, Dr. Marion Tichomirowa, Prof. Dr. Jochen Pilot, Dipl. Min. Heidrun Fuchs

http://www.mineral.tu-freiberg.de/techmin/achat_eng.html

41. *Geologists uncover agate origins, revamp theory of formation*

Roger K Pabian [Paleontological & gem expert] and Andrejs Zarins environmental quality geologist

http://csd.uni.edu/ResourceNotes/julyaugust1992.asp

42. *Black Opals of Lightning Ridge*

Elizabeth Smith

Simon & Schuster; Roseville NSW 2069 Australia; 1999

[pg 44-45]

43. *The Mineral Gallery*

The Olivine Group of Minerals
http://mineral.galleries.com/minerals/silicate/olivine.htm....

44. *Fossils & Fossilization* [p285]

Subject: Solubility of salts

JSTOR

45. *Chemical Weathering* – Carbonic acid

46. *Distance Learning and Educational Services*

Chem4Topic 26 [Carbonates]

http://www.distancelearning-tz.org/chem4topic26.htm

47. *Soil pH*

Lecture 12a Soil Chemistry
http://wwwsoils.agri.umn.edu/academics/classes/soil2125/lecture%20pp/

111a-chem.ppt

Silcrete & Mound Springs:

48. *Opal,*

The Opal Hut

http://www.opalhut.com.au/opal.htm

49. *North Dakota Rock & Mineral Sample - Silcrete*

North Dakota Geological Survey

http://www.state.nd.us/ndgs/rockandmineral/Silcrete.htm

50. *Mining Terms - Gannister*

The Durham Mining Museum

http://www.dmm.org.uk/books/terms_g.htm

51. *Silcrete Lab Analysis*

http://www.tec.army.mil/research/products/desert_guide/ls msheet/lssilc.htm

52. *Photocatalysis on Nano sized Semiconductors*

By Stephen Popielarski, 1998;

Website:
http://www.rpi.edu/dept/materials/COURSES/NANO/popi elarski/

53. *Advanced oxidation technologies; Photocatalytic treatment of wastewater.*

Author: Jian Chen (1997)

WAU dissertation No. 2309.

Source: http://library.wur.nl/wda/abstracts/ab2309.html

54. *A new Opal Model*
Mound Springs as Opal Source
Byron Deveson
Canberra 2006

55. *Smectite and Hot Spring Sinter Diagenesis*
Clays and Clay Minerals [Role of Bacteria].
Vol 55, No. 2, 2007 [Pages 189-197]
Kyle and Schroeder

56. *New World Encyclopedia*[23/02/2009]

Subject: Carbonatite

http://www.newworldencyclopedia.org/entry/Carbonatite

57. *Wikipedia*

Subject: Carbonatites

http://en.wikipedia.org/wiki/Carbonatite

58. *Opals Born in a Boom*

June 10, 2004

Source: The Black Opal Advocate

Opals Downunder [10/3/09].

59. *The Mystery Gem*

Page 1 [17/02/09].

http://www.contrafedpublishing.co.nz/QM/The+mystery+gem.html

60. *Investment Profile – Opal HiTech Ltd*

Martin Place Securities 20th March 2008 [pg 2].

Subject: Opal Exploration

61. *Nannobacteria and the Formation of Framboidal Pyrite: Textural Evidence.*

Robert L. Fock

Department of Geoscience, University of Texas, Austin Texas,

USA.

J. Earth Syst. Sci. 114, No. 3, June 2005, pp. 369-374.

62. *Columbia New Illustrated Encyclopedia*

Vol. 18 Pgs 5582-5583

Subject: Pyroxene

Columbia University Press, New York, NY 1978.

63. *A new understanding of the Groundwater Resources of the Great Artesian Basin.*

Occasional Paper, No.2, May 2001

By Prof. L.A. Endersbee AO FTSE

64. *Wikipedia*

Subjects: 1Pyroxene, 2 Nepheline, 3 Feldspathoids.

http://en.wikipedia.org

1 Sept 09

65. *Hutchinsons Encyclopedia*

Subject: Pyroxene

http://encyclopedia.farlex.com/Orthorhombic+pyroxene

Sept 09

66. *Geophysics*

http:hyperphysics.phy-
astr.gsu.edu/Hbase/Geophys/pyroxene.html

Subject: Pyroxene

1 Sept 09

67. *Mineral Galleries*

Subject: Nepheline   1 Sept 09

http://www.galleries.com/minerals/silicate/nephelin/nephelin
.htm

68. *Mineral Galleries*
Subject: Feldpathoids
http://www.galleries.com/minerals/silicate/feldspat.htm
1 Sept 09

69. *Answers. Com*
Subject: Feldspathoids
http://www.answers.com/topic/feldspathoids
1 Sept 09

70. *Lightning Ridge; Home of the Black Opal*
Gan Bruce
Subject: Silica concentration [102-104].
Macquarie Publications, 1983, Dubbo NSW Australia.

3 ORIGIN OF OPAL ORGANICS

Wood Chemistry

Decomposition chemistry of plants and animals is called Taphonomy.

The purpose of this chapter is to discover the organic chemicals and their functions in the creation of opal, and supporting evidence for the same.

3.0 Polymorphism

The Fenton reaction is involved in the decomposition of wood, phenols maybe involved in this process providing a link to Humic acids and humic substances. Note: Sawdust is classified as an ionic polyelectrolyte.

Another link between wood decomposition and the Fenton reaction is Lignin Peroxidase (LiP).

It is not only wood that can become opalized, there are a

number of materials that form opal polymorphs; these are:

- Wood

- Shells
- Glauberite
- Alunite

According to Elizabeth Smith polymorphs of Glauberite are formed when the crystals of Glauberite are washed out forming a cast in the clay which is later filled by Silicic acid. This method is probably the same for other inorganic materials that are opalized.

However it seems to me that the process is different for organic materials, which are replaced by the process of ion exchange with the opal electrolyte.

Cellulose can hydrate over time to form Glucose, it is possible that this glucose maybe converted to Carbonic acid, then replaced by the dissolved silica. The cellulose may be dissolved by the electrolyte.

3.1 Wood Decomposition Chemistry

Wood decomposes into the following:

- Cellulose
- Hemicellulose
- Lignin
- Extractives, which include:
  - Proteins

- Amino Acids and Nucleic Acids

Cellulose

Silica is sometimes involved in a process called Petrification, or opalization. This process is called polymorphism. Opalization is also referred to as silicification. It is now known by experimentation that wood can opalize within 7 to 10 years although it can take up to hundreds of years depending on conditions. The silica invades the tissue of wood through cracks or split surfaces of fallen wood and reaches a concentrated level whereupon it reacts with the cellulose and lignin of the wood tissue in the following manner:

$$C_2H_5(OH) + Si(OH)_4 \rightarrow$$

Ethanol + Silicic Acid

$$C_2H_5 - (OH)_2\text{-}Si(OH)_3$$

Next with additional Silica:

$$Si(OH)_3\, OSi(OH)_3 + H_2O \text{ Polymerizes to}$$

$$SiO_2 + 2H_2O \text{ Opaline Silica.}$$

It is reasonable to postulate that the reaction with ethanol could produce TEOS, if the conditions were anhydric. A little sulfuric acid could catalyze the esterification reaction to form TEOS. Water is produced through this reaction leading to hydrolysis followed by dehydrolysis.

Another explanation provided is that Polymerization of the

Silicic acid which has permeated the cells of the plant tissue, leads to dehydration (this may lead to second stage polymerization, author comment).

(Ref: 1, Rhynie Chert).

However if the ethanol becomes oxidized by water or other oxidizing agents it will degrade to Acetic acid, which reacts with cellulose to form cellulose acetate [$CH_3COOC_2H_5$]. This acetate represents an acetyl group CH_3CO plus an ethyl ester OC2H5 (Ethanoate). It is possible that the Acetate may combine with dissolved Silica to form TEOS [$Si(OC_2H_5)_4$]

Here are two possible organic routes to natural TEOS formation, if you look closely at the chemical formulas, you can see how nearly they resemble the TEOS formula. This does not rule out the possibility that a catalyst may also be involved..

Another reaction of cellulose is the hydration reaction by which cellulose degrades to Glucose.

It is not surprising then that Glucose can polymerize Silicic acid, as cellulose degrades to Glucose by the following reaction:

$$(C_6H_{12}O_4)n + H_2O \rightarrow C_6H_{12}O_6$$

Cellulose + water $\rightarrow$ Glucose

Fermentation of Glucose produces CO2

Another pathway for cellulose degradation is by acetate

fermentation by the following reaction:

CH_3COOH (cellulose) $= CH_4 + CO_2$ (acetate fermentation)

$CO_2 + 8H+ = CH_4 + 2H_2O$ (CO_2 reduction)

Methane (g) $= (CH_4)$.

Hemicellulose or Polyoses

- Xylans
- Mannans
- Glucans
- Glucomannans
- Galactans
- Pectins

Soluble in alkali and readily hydrolyzes in dilute acid to form sugar acids, xylans (polymerized pentose sugars), short chain polymers, glucomannans (hexose sugars).

Glucomannans

Glucomannans are a water soluble dietary fiber Konjac Glucomannan often described as a high molecular polysaccharide composed of Glucose & Mannose linked or bonded together with Acetyl groups. They are very similar to Pectin in function and structure. One popular variety is the Konjac Glucomannan (KGM). It is well known for its water gelling properties (Ref: 5).

Another source states that the acetate groups occur on carbon 6 on the main chain at regular intervals. It is the hydrolysis of this group that results in the formation of intermolecular hydrogen bonds initiating the gelling action.

Pectin

This Polysaccharide supports the cell walls of all plant tissues by acting as a cementing agent. The structure of Pectin consists of Mentholated esters of polygalacturonic acid. Contains 3x methyl ester groups (-COOCH$_3$) and 2x carboxyl groups (-COOH). It is also found in fruits and in the rinds of citrus. It is used as a gelling agent for fruits jellies and jams but must be used with sugar (sucrose) and with some citric acid for it to gel properly.

Lignin

Lignins are described as phenyl propane skeleton. They are cross linked polymers, the monomers of lignin vary but may contain Coniferyl alcohol (R=OCH$_3$) and Sinapyl alcohol (R=R=OCH$_3$) depending on species. Very similar to phenols (C$_6$H$_5$OH) and phenolic resins. The structure is complex and non-repeating. These are broken down by the free radical reactions, the Fenton reaction.

There is also a Lignin Peroxidase (LiP), have yet to investigate the origins of this substance. It could be specific to fungus or bacteria. *LiP* may not be a normal part of plant chemistry.

Put more simply and in general terms rather than being specific, Lignin is a polyphenol of high molecular weight, consisting of various types of aromatic carbon rings. Better to deal in generalities as different species have different alcohol types, but in general they are all polyphenols.

Extractives

Extractives include the above mentioned Proteins and biogenic acids, as well as:

- Terpenes & Terpenoids (used to produce terpentine).
- Fats (Carboxylic Acids)
- Waxes & gums (Esters?).
- Phenolic compounds, (Carbolic acid, or hydroxybenzene).
- Quinones
- Tannins
- Flavonoids

And also extractives from Foliage, Buds, and fruits.

Pyroligneous Acid

This substance is also known as Wood Vinegar and contains the following:

- Acetic acid
- Methanol
- Acetone
- Wood Oils & Tar

Chlorophyll

The active ingredient of photosynthesis found in plants where CO_2 is extracted from the surrounding air to produce sugars as an energy reserve for plants. There are two basic types of Chlorophyll:

- $C_{55}H_{72}MgN_4O_5$ Chlorophyll (a)
- $C_{55}H_{70}MgN_4O_6$ Chlorophyll (b)

There is a Metalloprotein in plants that acts as a catalyst in the photosynthesis process, it contains the metal cluster Mn_4Ca. The main function of this cluster is the activation of dioxygen.

Gelling Agents

Wood chemistry provides us with a number of gelling agents, these include:

- o Glucomannans
- o Pectin
- o Chitin/Glucosamine are contained in some forms of cellulose.
- o Hydroxyethylcellulose: cellulose derived polymer used as water gelling agent
- o Some polyamides; more likely to be animal derived than plant.

Acid in wood

> *"The cellulose polymer consists of linked glucose rings, usually about 10 000 subunits. An oxygen atom between the carbon atoms 1 and 4 joins the rings. These so-called glycoside linkages are easily broken (hydrolysed) in an acid environment, while fairly stable under neutral and alkaline conditions. The catalytic degradation is initiated by a hydronium ion (H3O+), binding to the bridging oxygen atom. This facilitates breaking of the bond between the oxygen and carbon atom 1. When a water molecule then binds to carbon atom 1, a hydronium ion is regenerated and can initiate a new hydrolysis reaction. The length of the cellulose chain will gradually decrease. Finally only small fragments of crystalline cellulose of about 200 glucose units will remain, and the tensile strength of the wood will have completely disappeared (Johansson 2000)."*

"The oxidation of sulfur to sulfuric acid described previously is probably catalyzed by the presence of iron ions. Another detrimental effect is that iron compounds catalyze oxidative degradation of cellulose (Johansson 2000). Tests have shown that wood in contact with rusting iron loses a considerable part of its tensile strength over time (Marian 1960)."[Ref: 7]

It is therefore not only the Lignin that is broken down by the Fenton reaction but also the Cellulose.

3.2 Case Studies of Polymorphism

It's time to look at some examples of polymorphism to gain a better understanding of the opal formation process and to begin to link together the clues which have been somewhat of a mystery in the past.

How is it that you can have opal form within clays, rocks, bone, shell and even wood. These mediums seem to be so different and yet opal can form in all of them. There must be some common ground that will help piece all these clues together in a way that understandable.

Common ground does exist, it is organic in nature and is derived from the breakdown of organic material. This in fact answers many questions for us, such as why opal has not formed everywhere in the Great Artesian Basin. Opal has only formed where there was opal dirt, moisture, some potassium and decomposing organic matter, and nuclearites. You see, it is the chemicals that result from the decomposition process that provides our elusive ionic electrolyte. The conversion of silica to opal requires a very strong electrolyte. That electrolyte could be a Humic or Fulvic acid which has a CEC of 1400 which is considered to be extraordinarily high. However it is almost certain that the ideal electrolyte is Carbonic acid.

3.3 Opalized shells

The same process prevails, the shells were not just calcium carbonate but also contained live sea animals before they were buried and died. The bodies of the shell creatures decomposed once again providing the amino acids which are essential as a source of Carboxylic acids from which Carbonic acid is derived and the Amines from which the ammonia is derived. Carbonic acid can also be formed from the $CaCO_3$ from which the shell is composed.

3. 4 Opalized wood

It appears that the iron in the environment of the buried wood enters into the Fenton reaction breaking down the Lignin and the Cellulose of the wood. The Fenton reaction also has an effect on the clay, dissolving first it would seem the Al content of the clay allowing the Silica component to dissolve in the presence of the water and Hydroxyl ions generated by the Fenton reaction. It is more than likely that iron oxide generated interacts with Al to form Al_2O_3 and iron.

3. 5 Lake Tateyama – Opalized wood [Japan]

Opalized wood has been found in the vicinity of many volcanic regions, however in recent times it has been discovered to be happening currently in a hot water spring at Lake Tateyama (Toyama Prefecture). This lake has high silica content and is said to readily precipitate silica spheres and deposits of opal. [Ref: 9]

Australian Farmers have noticed that in very sandy soils fence posts turn to opal. A farmer put up a fence made of wooden posts and wire. Five years later he was taking the fence down only to find that the bottom part of his fence posts had turned to opal, the soil around the bottom of the post was dark indicating humic matter in the soil.

3. 6 Petrified wood

"Wood must first be covered with such agents as volcanic ash, volcanic lava flow, volcanic mud-flows, sediments in lakes and swamps or material washed in by violent floods – by any means which would exclude oxygen and thus prevent decay. A number of mineral substances (such as calcite, pyrite, and marcasite) can cause petrification, but by far the most common is silica. Solutions of silica dissolved in ground water infiltrate the buried wood and through a complex chemical process are precipitated and left in the individual plant cells. Here the silica may take a variety of forms, it may be agate, jasper, chalcedony or opal. The beautiful and varied colors of petrified wood are caused by the presence of other minerals that enter the wood in solution with the silica. Iron oxide stains the wood orange, rust, red or yellow. Manganese oxide produces blues, blacks or purple."[Ref: 6, Hamilton Hicks, Petrified wood].

Hamilton Hicks of Greenwood Connecticut, invented a method for making Petrified wood. It consists of a cocktail of Sodium Silicate and natural spring or volcanic mineral water which has a high content of calcium, magnesium, manganese and other metal salts, and citric or malic acid is capable of rapidly petrifying wood. [United States Patent Number 4,612,050, September 16,1986]

But only if done correctly, the gelling agent [acid] must be in the correct proportion so that gelling does not occur until after the solution has penetrated into the wood [Ref: 6].

3. 7 Burnt Wood

Burnt wood form two distinct and significant chemicals:

- Charcoal is almost pure Carbon, and
- Wood Ash, the grey left over ash after burning, contains:
  - $KnaO$, CaO, MgO, P_2O_5 [Phosphorus] & SiO_2

Natural Oxides, See also: CH 4.

Animal Decay Products

3. 8 Decomposition Chemistry & Opal Formation

What is the link between the formation of opal in the ground; opalized wood; opalized shells; opalized animal bones and other types of opalized organics. These are the nuclearites that attract the silica ions in the ion exchange process [see CH 5 Len Cram]

There must be a common link between all these seemingly different types of opalization. What is the link? What is the commonality between the decomposition of animals and the decomposition of plants.

The link is the electrolyte. The same electrolyte is responsible for all these opalization events. As the animals or plants decomposed they produce the electrolyte that alters the silica into opal. In the case of the organics another process called ion exchange is also involved. It is the process of ion exchange not replacement that causes the ions in the organics to be exchanged for the dissolved silica. The silica then becomes polymerized by the electrolyte possibly with the help of a catalyst. Replacement is an antiquated generalized term used to explain in a non specific manner the scientific process we now know as ion exchange.

The organic components chemicals are provided by the decay of Organic matter, these being the ions of C, H, N, O & S. The decay of plants and animals provide these organic chemicals.

When an animal or plant dies it begins to decay; decompose, to breakdown. In the decomposition of animals the body begins to digest itself, the stomach acids eat through the stomach lining and start to digest the body proteins releasing the amino acids which are released from their bonds and react to form both ammonia salts and Nitric Acid.

Two of the major sources of decomposition are:

- The Flesh
- Fat

Body flesh consists mainly of the muscles and tissues, it is almost entirely protein. Protein contains the amino acids. It is the amino acids that contain both the carboxylic group and the amine group NH2. It is from the decay of the amine group that Nitric Acid forms.

Body fats are produced by the body's storage mechanism of carbohydrates and sugars which theoretically can be decomposed back to sugars providing a fermentation source from which the ethanol results.

3. 9 Decomposition of Organic Matter

The decay of organic matter is referred to as one of the following four terms:

- Oxidation
- Metabolism
- Degradation
- Mineralization

For the process of aerobic respiration to occur, there must be sufficient dissolved oxygen with bacteria which oxidize the NH_4 producing a nitrite (NO_2) and is then converted to nitrate (NO_3) in a process called nitrification.

It needs to be understood that under different conditions and with different minerals present, that the method of decomposition will be differ accordingly. Methods of decomposition include:

1. Oxygen Reduction
2. Manganese Reduction
3. Nitrate Reduction by Denitrification
4. Dissimilatory Nitrate reduction to Ammonia
5. Iron Oxyhydroxide Reduction
6. Sulfate Reduction

These reactions are influenced by the amount of dissolved oxygen present. In reactions 2-5 occur under sub-oxic conditions but sulfate reduction can only occur under anoxic conditions. As dissolved oxygen is depleted it creates the right conditions for redox reactions, which are inhibited by oxygen, but thrive in an oxygen depleted environment.

3.10 Proteins, Amino Acids & Enzymes

It is the protein [Amides] that provides the amino acids from which the ammonia and the carboxylic acids are released through the fermentation process. Yeast alone will probably not provide enough amino acids for the production of ammonia and amines.

The protein in the ground that is the major source of amino acids and ammonia for opal formation comes from dead animals and plants. As plant and animal bodies decay in the ground they are providing the building blocks for opal formation.

Another possible link in the formation of opal may well be nitric acid, which is one of the final products from the decay of organic matter. One of the properties of Nitric Acid is that is very reactive, so does not remain as nitric acid for long but reacts with other elements in the soil resulting other compounds being formed.

How does nitric acid form in the soil (nitrogen fixation)?

The degradation of protein forms the following acids:

- Sulfuric
- Phosphoric
- Nitric acids

3.11 Decarboxylation

Amides are easily hydrolyzed to carboxylic acids and ammonia in the presence of mineral acids or strong metallic hydroxides. Proteins are amides.

3.12 H2O2 & the Ruff Degradation

Hydrogen peroxide acts as a mild oxidant in the presence of Iron salts, catalyzing the oxidative decarboxylation of a-hydroxyacids to give CO2 and the shorter aldehyde. This is a well known chemical reaction in the sugar industry which refers to it as the "Ruff degradation" [Decarboxylation].

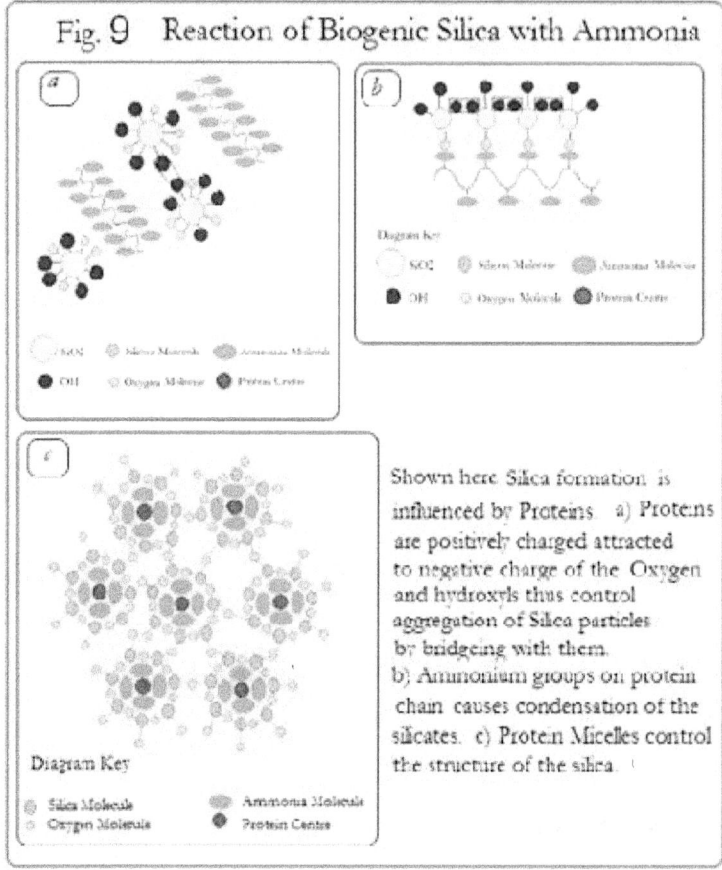

Fig. 9   Reaction of Biogenic Silica with Ammonia

Shown here Silica formation is influenced by Proteins. a) Proteins are positively charged attracted to negative charge of the Oxygen and hydroxyls thus control aggregation of Silica particles by bridging with them. b) Ammonium groups on protein chain causes condensation of the silicates. c) Protein Micelles control the structure of the silica.

[Refer: ChemBioChem 2003, 3, 1-3 Biogenic Silica Patterning, P.J Lopez & T. Coradin].

Figure 10 Amino Acid Structure

3.13 Amino Acids & decay products

Another major source of carboxylic acids is the amino acids, which also contain the amine group NH_2.

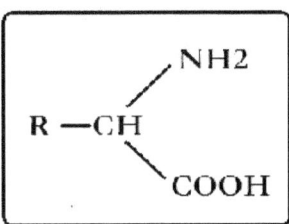

It should be obvious from the above that both carboxylic acid and polyamines are released by the decay of animal flesh and vegetable matter. The breakdown of organic matter is known to form the organic acid, HNO_3 or Nitric Acid (could be a major clue -- is this a catalyst for the formation of TEOS). This then allows these chemicals to interact with the available soluble silica along with ethanol.

Note that the NH_2 ions are amine ions, thus a polyamine is a chemical that contains more than one NH_2 group in its structure.

Amines can be prepared by reacting ammonia with alkyl halogenides.

Amino acids contain both Amines and carboxylic acids.

Carboxylic acids + alcohols form esters, very slowly.

Carboxylic acids react with ammonia to yield ammonium salts.

One particular Amino Acid, an AHA (Alpha Hydroxy Amino) is Glycolic Acid but it is also called "acetic acid hydroxy". It is very soluble in Water, Ethanol, Acetone, Acetic acid, Ethyl acetate and Methanol.

Figure 11. Alpha Hydroxy Amino Acids [AHA's].

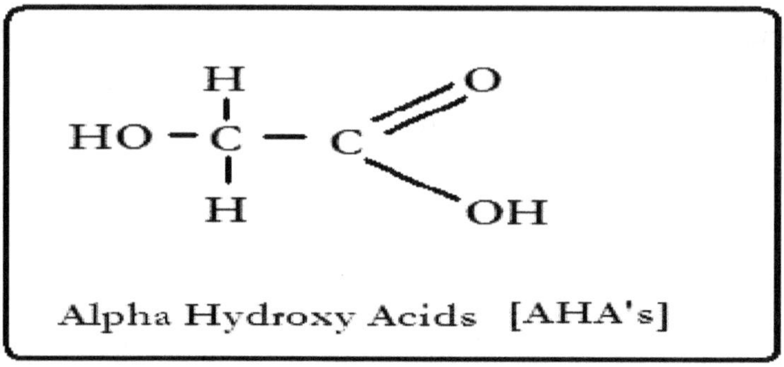

Alpha Hydroxy Acids [AHA's]

The degradation products of Glycolic Acid are:

- With Zinc & H_2SO_4 = Acetic acid
- With Ferrous salts converted to glyoxylic acid & H_2O by H_2O_2
- With Ferric salts conversion by H_2O_2 to:
    - Oxalic acid
    - Formic acid
    - $CO_2 + H_2O$ (gives H_2CO_3 Carbonic acid).

Glycolic Acid can also self estrify by heating to produce Carboxymethylhydroxyacetate, glycoside or polyglycolic acid. It can also speed-up the breakdown of amines.

Another organic possibility exists. Another organic electrolyte derived from the decomposition of both Animal and Plant decomposition. This common feature between plant and animal is the chemical cocktail that we call Amino Acids.

Aminoethanol

Aminoethanol is an organic compound of the formula $NH_2C_2H_4OH$. It is also known as monoethanolamine, ethanolamine, 2-aminoethanol and colamine. Aminoethanol is a colourless oil with a boiling point of 318 K. It is a strong base, volatile in steam and absorbs both carbon dioxide and water.

Water Soluble Amino Acids

It is the nature of amino acids to be generally soluble in water because of their Amine and carboxyl groups. Therefore it is probably easier to begin using Amino acids rather than proteins.

Amines

Amino acids contain Amines which are an ammonia derivative having the NH2 group as part of their structure. Amines have only one NH2 group whereas polyamines have more than one CH_2NH_2 group. Polyamines are known to dissolve silica and accelerate Silicic acid polymerization, as is $Al(OH)_3$ Aluminium Hydroxide.

The ammonia group NH_3 which is known to be responsible for sphere formation and sphere size control.

Shape control can be achieved by using synthetic organogelators, polypeptides, and polyamines.

Amines can be broken down by an enzyme driven oxidation reaction known as deamination. [See Amine Oxidases under Enzymes].

Carboxylic Acids [See Annex G]
Carboxylic Acids are the main organic acids in natural systems, together with amines they form Amino Acids which link together to form proteins [amides]. Many other organic compounds contain carboxylic acids. These natural acids are more effective in dissolving silica than the inorganic acids, which tend to be inefficient at this task.

Enzymes & their functions

3.14 General Information

Enzymes are organic catalysts which are responsible for initiating many different organic chemical reactions. Different enzyme types have different functions, such as Proteolytic enzymes. These are a class of digestive enzymes, which include lipases, and amylases. They are needed to digest the protein in food. Only small amounts of the animal-based proteolytic enzymes trypsin and chymotrypsin are found in the diet; however, the pancreas can synthesize these enzymes. The plant-based proteolytic enzyme bromelain comes from pineapples and papain comes from unripe papayas. All of these enzymes are available as supplements. Chicken livers may also be a source of some of these digestive (proteolytic) enzymes.

Proteolytic enzymes break peptide bonds.

3.15 Diatoms & Sponges

It is well documented that Diatoms produce opal. Latest research indicates that the Diatom is capable of dissolving silica and using it to form their shells.

Diatoms achieve this by using Proteolytic Enzymes in an acidic environment. Enzymes are related to Proteins and amino acids. Some amino acids contain amines, and some of them contain ammonia. Ammonia causes silica to form uniform spheres, which we know are needed to produce opal. (Polycatonic peptides also direct nanosphere formation)

It seems that diatoms may also contain phosphatids (^phosphatidylserine) which are a form of phosphorus and is responsible for fluorescence.

Polycatonic peptides are contained in Amino Acids.

Digestive Enzymes types & function:

Protease breaks down protein (amino acids)

Lipase breaks down fats (glycerol + Fatty acids)

Amylase breaks down starch to simple sugars.

3.16 Carboxylase

Carboxylase (Decarboxylase?) One of the most important enzymes in alcohol production. It is secreted by yeasts and its action is to promote the **decarboxylation** (removal of carbon) of pyruvic acid to acetaldehyde. In the process Carbon Dioxide is given off. Unlike most of the reactions involved in alcohol production this one is irreversible.

3.17 Enzyme Catalyzed Decarboxylation

This may sound simple but it must be kept in mind that this process is Species Specific, you must use the right Decarboxylase with the right Amino Acids, and i.e. Lysine requires lysine decarboxylase.

Decarboxylation of Amino Acids (Amines NH_2 + Carboxylic acid $-COOH$) should result in the release of NH_2 and H ions plus CO_2. The Metalloenzyme (contains ZINC) Carbonic Anhydrase should yield large quantities of Carbonic Acid, enough to dissolve silica for Opal gel formation. Note that the Carbonic Anhydrase may also be species specific i.e. Bovine Carbonic Anhydrase.

Another of the properties of Carbonic acid is that of a hydrating agent, that is a water binding agent, binds water not only to the Al3+ but also to the Silicic acid forming the OPAL GEL.

Note: Decarboxylation can also be achieved by the "Ruff Degradation".

3.18 Clay Catalyzed Decarboxylation

Smectite Montmorillonite clay] is a possible catalyst for decarboxylation reactions and is thought to be intimately linked to the petroleum generation process. Smectite Catalyses the decarboxylation of fatty acids and the production of long chain alkanes in a liquid hydrocarbon cracking process, similar to that used in the commercial production of petroleum. The association with oil generation extends even further with the discovery that certain carboxylic acids used as a precursor to oil formation have the ability to dissolve certain carbonates, and aluminosilicates which provides the basins for the oil to accumulate within. The organic matter in the clay possibly acts as a catalyst in clay dissolution. [Ref 36; pgs 22, 35]

3.19 The Claisen Condensation

The Claisen condensation involves the breakdown of amino acids involving a decarboxylation process which is catalyzed by the ethoxide ion, most notably the silicon ethoxide ion. This indicates that silica present in the decay of organic matter has been dissolved in order to combine with ethanol forming silicon ethoxide, TEOS. See CH 4.

3.20 Carbonic anhydrase

It is found in red blood cells where it catalyzes the reaction:

$$CO_2 + H_2O <\text{-}> H_2CO_3$$

One molecule of carbonic anhydrase can process one million molecules of CO_2 each second.

3.21 Glutamates are highly soluble in water; result from the degradation of Glutamine. Glutamine makes up to 60% of the body. One common glutamate used in cooking is Monosodium Glutamate (MSG) which contains $C_5H_8NO_4Na$ Carbon; Hydrogen; Nitrogen & Oxygen as NITRITE; & Sodium.

3.22 Glutamate dehydrogenase

Oxidatively deaminates glutamate to form <u>ammonia.</u>

3.23 Lysosomes [Digestive Enzymes]

These are often called the Garbage disposal system of cells; as they are like the digestive system of the cells. They contain up to 40 enzymes which are capable of breaking down all the major classes of molecules.

Enzymes in the Lysosomes are released upon death beginning the autolysis process which together with bacterial action aids in the decomposition of organic matter.

Lysosomes are produced in the *Golgi apparatus* of the cell. If you research biosilicification you may run across this term.

These digestive enzymes are a group of Hydrolases or Hydrolytic enzymes. These include all the Nucleases; Proteases; Lipases; and Amylases. They function best in slightly acidic conditions around a pH of 5 and are thus called acid hydrolases, or hydrolytic acids. They catalyze the hydrolytic cleavage of C-C, C-N, and C-O and other bonds of Amino Acids and proteins, this also includes the Phosphoric anhydride bonds. This is a very large group of enzymes which includes Monoamine Oxidases [MOA] and Superoxide Dismutase [SOD].

Recent studies in Biotechnology has opened new routes into the fabrication of nanotechnology by utilizing proteins from marine sponges and other organisms such as diatoms, which are now being called Silicateins. These Silicateins however are really ENZYMES which have been found to be related to a family of Hydrolases. These enzymes are being used as catalysts in the polycondensation of Alkoxides to form silica spheres, and Siloxane [Ref: 34].

MetalloProteins & Metalloenzymes

3.24 General Information

[Source Wikipedia, Web based]
Some proteins contain Iron & Sulfur clusters which act as catalysts for certain reactions often providing an electrical pathway for the reaction.

> *"Protein containing non-heme iron centers are widespread in nature. They perform a broad range of functions, particularly activating dioxygen for the oxidation of various substrates [Ref :39]."*

The most common metalloproteins containing metal clusters are:

- Rieske proteins
- Cysteine
- Polymetallic proteins [Ferredoxins]

The enzymes [metalloenzymes] within these proteins that carry the metallic clusters are:

- Oxygenases
- NAHD dehydrogenase; Co-Enzyme Q-cytochrome
- Reductase and
- Nitrogenase
- GAO
- MMOH [Methane Monooxygenase Hydroxylase]
- Ruberythrin
- Rubredoxin
- Hermerythrin
- Ferroxidase

These enzymes contain one or more of the transition metals with the most common being Iron and copper. Some of the Iron metalloenzymes also contain Sulfur. It is only the non-heme iron metalloenzymes that are of interest as they can be used as iron source for Fenton reactions and may be catalysts in the polymerization of silica.

The main function of these metalloenzymes is the activation of Dioxygen [O_2] the oxygen molecule with which we are most familiar.

> "The Hydroxylase component (MMOH) of the soluble methane monooxygenase (sMMO) enzyme is a non-heme diiron protein that catalyzes the hydroxylation of C-H bonds by dioxygen in a wide range of substrates, including methane [Ref: 38]."

3.25 Methane Mono-oxygenase Hydroxylase [MMOH$_{red}$]

Route to TEOS formation:

> "Hydroxylase (MMOHred) contains a diiron cluster [Fe111-- Fe111][Fe111-- Fe11][Fe11--Fe11] the diferrous state of hydroxylase (MMOHred) is the only one capable of reacting with dioxygen & initiating the catalytic cycle polymerizing *Silicic acid tetra ethyl ester." [Ref: *Reward for Missing Ref See last page of book for Reward details]

The unique ability of the Diiron species seems to be as a catalyst that activates dioxygen. In the opinion of the author it is not just MMOH that has this function but diiron species as a whole including FeS_2 which in water produces the diiron species. An important property of iron clusters is that ammonia dissociates when attaching to them. [Ref 41].

For plant Metalloenzymes see under Chlorophyll in previous section.

3.26 Amine Oxidases (AO) [ENZYME]

This group of Enzymes is present in both Prokaryotes & Eukaryotes.

Copper containing Amine Oxidases are found in Bacteria, Fungi, Plants and Animals. The oxidizes that contain a copper core catalyze the oxidative deamination of primary amines by dioxygen [Refer Diiron hydroxylase MMOHred (Fenton Reaction)]to form:

- 1st Aldehydes, then
- Ammonia and
- H_2O_2.

One of these enzymes is known as AGAO. The reaction is as follows:

$$RCH_2NH_2 + H_2O + O_2 = RCHO + NH_3 + H_2O_2$$

One of the active ingredients in copper amine oxidases is Topaquinone (TPQ). The copper ion being co-ordinated with three histidine residues and two water molecules. Besides ammonia & ammonia salts poly-l-histidine is known to enable the formation of silica spheres 150-200nm in size.

The Role of Bacteria

3.27 General Information

In considering the chemistry of body decomposition one cannot ignore the role of bacteria which help to breakdown organic remains. Bacteria may be involved in putrification but it is definitely involved in the fermentation process. The bacteria are the source of the enzymes that cause the body sugars to ferment. One particular enzyme that bacteria release is called Zymase which is the same enzyme released by yeasts in the fermentation of sugars.

There are some scientists who believe there is a link between bacteria and opal formation, having found bacteria in the opal environment.

It is my belief that the enzyme (zymase) released by these bacteria that catalyze not just the fermentation of glucose (C6H12O6) but that it catalyses the formation of TEOS by combining the resultant ethanol with the monosilicic acid. Both plant and animal matter contain fermentable sugars of which glucose is the most significant.

Ethanol also acts as a solvent for TEOS. Acetic acid [vinegar] may be a solvent for Monosilicic acid; additionally Acetone may also be such a solvent.

3.28 Bacterial in the Opal Environment

Bacteria are known to be part of the decomposition process which nature uses to recycle dead organic matter. Bacteria have been found in the presence of opal. The question is do they play a role in opal formation?

Recent evidence suggests that bacteria may play an essential role in opal formation. Bacteria excrete chemicals that create optimum conditions for opal creation.

> *"Siliceous sinters typically form in spring fluids supersaturated with respect to amorphous silica (they can also form from evaporating near-saturated spring fluids). Opal-A formation begins by nucleation in a colloidal suspension that then deposits out of solution (Rimstidt and Cole, 1983). Microbial cells often act as site templates for opal-A nucleation and subsequent sinter formation can then continue due to the lower surface energy barrier towards precipitation."* [Kyle & Schroeder, pg 189].

The above evidence gives bacteria the essential role of nucleating agent which is of paramount importance in the genesis of opal. Other researchers had assumed that the nucleating agents for opal were inorganic chemicals.

Some of the known types of bacteria that inhabit spring waters today are:

- Aquifex
- Thermotoga

Bacteria of this type are referred to as hyperthermophilic microbes, which tolerate the alkaline and heated conditions of these waters.

Bacteria are almost everywhere, and would certainly be in the decay products of bone, wood and shell. Now it is a real probability that they would be responsible for nucleation in the process of polymorphism.

Processes: Associated with decomposition
- Putrefaction
- Fermentation [Butyric]
- Saponification (Soap Mummy) ; Emulsification
- Esterification [See Ch. 4 Esterification].

3.29 Putrefaction

Produces ammonia, and foul smelling gases. Putrefaction happens with both plant and animal decomposition.

Fermentation is a large part of both animal and plant decay producing the alcohol ethanol in the process. Nitrogen is also produced by the fermentation of animal matter.

Another aspect of body decomposition that may possibly be involved is the soap mummy. This may occur when temperatures are low and decomposition does not occur in the normal manner. It implies the availability of caustics which cause the body fat to saponify.

3.30 Methyl Mercaptans

Methyl Mercaptans [Thiols] is a colorless gas with a smell like rotten cabbage. It is a natural substance found in the blood, brain, and other tissues of people and animals. It is released from animal feces. It occurs naturally in certain foods, such as some nuts and cheese. The chemical formula for methyl Mercaptans is CH_3SH. Methyl mercaptan is released from decaying organic matter in marshes and is present in the natural gas of certain regions in the United States, in coal tar, and in some crude oils.

TABLE 9 - **The main products of Putrification**

Ammonia NH_3	Released from Putricine & Cadaverine
Phenol	Carbolic acid (C_5H_6OH) an alcohol.
Putricine	polyamine; $NH_2(CH_2)_4NH_2$ (1,4-diaminobutane or butanediamine) formed by and having the smell of rotting flesh Note that putrescine is synthesized by *healthy* living cells by the action of ornithine decarboxylase
Cadaverine	Cadaverine is a toxic di-amine with the formula $NH_2 (CH_2)_5NH_2$. Cadaverine is the decarboxylation product of the amino acid lysine.
Sepsin	"endotoxin," or lipopolysaccharide (LPS).
Methyl mercaptan	formula for methyl mercapatan is CH_3SH. Methyl mercaptan is released from decaying organic matter

Examples of Thiols [Mercaptans] are:

- Methanethiol – CH_3SH
- Ethanethiol – C_2H_5SH
- Coenzyme A
- Lipoamide
- Glutathione
- Cysteine

The functional group of the amino acid Cysteine is the thiol group which has an important function in biological systems.

The functional group of thiols are the Sulfur atom and the Hydrogen atom (-SH) which are referred to as the thiol group or the sulfhydryl group. Some amino acids contain the sulfhydryl group. Sulfur and yeasts can also form thiols.

The Sulfur ions of a thiol are very nucleophilic, by itself is acidic but in the presence of a base, a thiolate anion is formed which is a very powerful nucleophile. When combined with powerful reagents like Sodium Hypochlorite the result is Sulfonic Acids [RSO_3H].

3.31 Sulfonic Acids

Can be prepared as stated above, or can be prepared from Amino acids. Sulfonic acids are used as catalysts in alkylation, condensation reactions and esterification reactions. They are often used in fabric dyes for colour fastness. However they also form very strong cationic exchange resins with silica, and are widely used in industry for this purpose.

The salts or esters of Sulfonic acids are known as Sulfonates, available as powders they are used extensively in the detergent industry. As ion exchange is believed to be an important process in opal formation, sulfonic acids may be the natural ion [cation] exchange electrolyte. As shown above Sulfonic acids could well form in the opal environment, as opal formation is reliant on the decay of organic matter to produce the organic components that interact with silica to produce the opal.

One particular organic dye is 2:2 BDSA; 4,4'-Diamino Diphenyl 2,2' Disulfonic Acid: This Amino acid dye product has 2 amine groups and 2 Sulfonic acid groups; therefore could be the electrolyte and could replace need for Ammonia.

3.32 The main products of Fermentation are:

- Butyric Acid ($CH_3CH_2CH_2COOH$) [Butyric Fermentation]
- Ethanol (Ethyl Alcohol) C_2H_5OH
- Acetic Acid (Vinegar) CH_3COOH
- Nitrogen/ Nitric acid (from animal decay).

Ethanol has an affinity for silica reacting and combining with it to form TEOS, however on heating TEOS the ethanol component is converted to its gas state, separating from the TEOS to leave $SiO_2 + H_2O$, this is called hydrolysis.

Ethanol can also be produced by fermenting molasses. The enzyme Zymase is released by the yeast. They catalyze the sugars causing fermentation which produces ethanol and carbon dioxide.

$$C_6H_{12}O_6 \rightarrow 2C_2H_5OH + \mathbf{2CO_2}$$

Acetic Acid is produced when fermentation is incomplete, commonly known as vinegar, cheap wines that have not completed the fermentation process properly will taste more like vinegar than wine.

Both Ethanol and Ammonia become gaseous at very low temperatures. For Ammonia this temperature is 35°C and both of these products create a cooling effect upon evaporation and condensation.

To produce artificial opal the mixture is heated to 35°C for a short time and then cooled slowly. It is at a temperature range of 28-20°C that crystals begin to form, this has been observed using a 50X magnification.

This crystal formation is in fact Opal Gel. Once this process has begun, given that conditions remain fairly constant, it will continue moving from a hydrolysis phase to a condensation polymerization phase. When the bulk of the mixture has crystallized, the remaining fluid is drained off so that dehydration polymerization can complete the process. After this has occurred, the Opal Gel will need to be dried, in order for the Opal Gel to harden into Opal. The best way to do this is probably to use an opal dirt mold under the summer sun. When the mold is baked by the sun, the opal should have reached normal hardness, after-all this is nature's process.

The crystals that form the opal gel are in fact Christobalite which is a natural Photonic Crystal.

Yeasts that combine with or are contaminated with sulfur tend to produce Thiols [Mercaptans]. [Ch. 4 Fermentation].

3.33 Saponification

Another process which often happens in the decomposition of dead animals is the saponification process. The body has tissue that is called Adipose tissue which is connective tissue that is used by the body as the major storage site for fat, in the form of triglycerides.

When a body is buried with little or no oxygen it will undergo adipocere formation. The bacteria convert body fats to adipocere are anaerobic. These bacteria digest the body fat excreting adipocere plus ammonial gases.

Much if not all body tissue will go through several decomposition stages, possibly ending with transformation into adipocere of most if not all of the fatty tissue. Other body tissue can also undergo this process but it is not fully understood as yet. Adipocere is also known as:

- Grave wax
- Mortuary Fat
- Saponified tissue
- Soap mummies

Alkaline sources also act upon fat tissue in much the same way as lye acts upon vegetable oil/fat. It hydrolyses the oil/fat turning it into a type of soap. The fats involved in this process are the fatty acids or triglycerides:

- Myristic
- Plamitic
- Palmitoleic
- Stearic (Lard?)
- Oleic

- Linoleic acids

NOTE: Although fatty acids are Carboxylic acids, some also contain:

- Glycolic acid/Glycosides
- Amines

Saponification is a form of Emulsification; a mixture of oil & water requiring an emulsifier to bond them together. In the case of saponification that emulsifying agent is an hydroxide (normally sodium or potassium). With water-in-oil emulsions it is a detergent.

The remaining liquid from soap making is called GLYCEROL.

3.34 Water Absorbents

Water absorbents in the opal environment may allow reactions to take place that would normally be hindered by the presence of water [i.e. Esterification].

Some AMIDES & polyamides have very good water absorbent properties. SLUSH Powder used by magicians and also in baby nappies.

You will find in Magician Supply shops.

Water absorbing agents are:

- Polyamides
- Polyacrylates
- Glucomannan
- Pectin
- Hydroxyethylcellulose (HEC)

Hydrating Agents

- AHA's Alpha Hydroxy Aminos Acids Ie Glycolic Acid (see Amino Acids).

- Carbonic Acid

Buried Organic Matter [Diagenesis]

3.35 General Information

There is no disputing the fact that opal forms underground. It is also well known that the organic chemicals required for its formation are derived from the decay process. The question that remains is "What happens when organic matter is buried in clay stone"?

When organic matter is buried in clay stone [opal dirt] it begins to decay releasing organic chemicals including water, as protein, the largest part of animals is approx. 70% water. The water begins to wet the clay and hydrolysis with the clay begins. Clay hydrolysis will release excess silica as dissolved silica and water. The clay will eventually transform to a different type of clay. Ammonia will also be released, as will ethanol. These are two of the organics responsible for opal genesis. Bacteria are also present from two sources, both the clay and the decaying organic material in which it multiplies. It is uncertain if aerobic bacteria can survive in these conditions, if not other types of bacteria will flourish.

Bacteria are known to accelerate the decay process and have an affinity for dissolved silica. Bacteria can excrete surface active chemicals directly onto the silica spheres expediting the opal forming process.

Deveson states that montmorillonite which is present in high amounts in clay stone can pressurize the opalizing fluids and provide a medium for ultra filtration and dialysis. The mere burial of organic decaying material in opal dirt then provides all the chemicals and conditions required for opal formation. [Supporting evidence: Opal Miners cat; Australian farmer's fence posts.

Byron Deveson's theory seems to be well founded, and one of the clearest accounts for opal formation, yet it seems that it only explains relatively small deposits of opal. Some of the opal shelves in Australia run for hundreds of kilometers. This would require a massive amount of organics and dissolved silica spheres 'en masse'. Another thing that this theory does not explain is 'Polymorphism' [the transformation of wood, bones and shell to precious opal]. Neither does it explain Virgin Valley opal which is a type of polymorphism. Virgin Valley opal starts by volcanic activity but as the trees became covered in ash then the process must be considered sedimentary in nature. It is highly unlikely that opal formed instantly, which would suggest that it was volcanic opal. The heat of the volcanic ash would have an accelerating affect on chemical processes, but once the heat dissipated then it must be considered to be sedimentary in nature. This process of polymorphism is related to petrification [See: Hamilton Hicks].

This is not an isolated case as many opalized bones have been found, including many species of dinosaurs [CH 5; ref: 2]. Seashells have also been found opalized, remember they once contained living things, therefore this is once again the result of decay.

Australian Sedimentary opal formed after the mass burial of organic matter within the layers of opal clay. Evidence of this is the enormous amount of opalized fossils that have been dug up in opal mining areas such as Lightning Ridge. Wood is also amongst the items that have fossilized as opal. Virgin Valley's opal is almost entirely formed from forests that have been felled by volcanic activity and then covered by volcanic ash. These burial conditions are anoxic, lacking oxygen, creating the ideal environment for silica to dissolve. Australian farmers have also reported that the bottoms of fence posts which have been placed into sandy soils having been converted to opal in as little as 5 years.

This process is also evidenced by the enormous array of opalized fossils that have been found around the world and at Lightning Ridge. The range of fossils includes Pine cones, huge & small dinosaurs [plesiosaurs, pterosaurs, sauropods, theropods, and ornithopods], platypuses, turtles, birds, protozoans, crinoids, mussels, snails, whelks, bony fish, sharks, crocodiles, lungfish, and yabbie remnants. [Elizabeth Smith, 54].

Most opal polymorphs with very few exceptions were once living things.

It is quite possible that this process could have happened in as little as one or two years. In warm to hot climates bodies tend to decay within 12 months. These polymorphs are also describes as nuclearites, referring to wood, bone and shell. These all contain a certain amount of Biogenic Silica, which seems to be the opal precursor in nature. It is easily soluble silica with a very small particle size. This dissolved silica together with Ethanol and the catalyst $MMOH_{red}$ could well form TEOS.

Ikaite

3.36 Basic Information

The references above lend support to the probable formation of Ikaite in prehistoric Australia. Recently a new link to polymorphism was discovered, it is a form of Calcium Carbonate called Ikaite. In this form it is crystalline variety that contains six molecules of water, referred to as a hexahydrate of Calcium Carbonate [$CaCO_3.6H_2O$]. This mineral substance is formed only in near freezing waters, below 10°C. Opal pineapples discovered at White Cliffs have an identical surface structure to the Ikaite. This, along with evidence below, is an indication that Ikaite is an ion exchanger with silica, in particular opal spheres.

Pseudomorphs believed to have been created in a cold water environment with Ikaite as the ion exchange electrolyte are:

- Glendonite, after type locality, Glendon, New South Wales, Australia.
- Thinolite, (Gr. Thinos = shore) found in the tufa of Mono Lake, California, USA.
- Jarrowite, Jarrow, Northumberland, UK.
- Fundylite, Bay of Fundy, Nova Scotia, Canada
- Gersternkorner, (Ger. = Barleycorn)
- Gennoishi, (Jp. = hammerstones),
- Molekryds, (Dan. = Mole Cross), Mors Island, Jutland, Denmark
- Pseudogaylussite (from semblance to Gaylussite)

White Sea hornlets, White Sea and Kola peninsula.

Ikaite forms in cold water but may also need trace chemicals which act as nucleation inhibitors for anhydrous calcium carbonate [Calcite inhibitors], such as the phosphate ion. Oxidizing Methane may be a nucleating inhibitor. Ikaite has been seen to form at Hokkaido saline springs in the winter time, but can be found almost anywhere in the world where there is near freezing waters.

Large columns of Ikaite exist in "The Ikka Column Garden" SW of Greenland. It is named after the Ikka Fjord. It is currently being studied by 'The Ikka Project'.

[Eckert, 50-51; Elizabeth Smith, 55,58; Wikipedia]

Burial of organic matter causes an anoxic condition, which means there is a depletion of oxygen. Under these anaerobic conditions you would expect anaerobic microbes, and bacteria to invade the organic matter, helping to accelerate the decomposition of this organic matter. In an aluminosilicate environment these conditions cause the development of Silica-organic-acid complexes which cause the rapid dissolution of silica. The decaying process often releases ethanol by way of decarboxylation, which may combine with dissolved silica to form Silicon ethoxide. Ethoxides seem to accelerate the decay process releasing more ethanol and ammonia would also be expected to be released as the organics breakdown. Oxides which are naturally occurring soil catalysts act as catalysts for ethoxide formation. Potassium tends to be a promoter of these oxide reactions whereas Lithium is an inhibitor. [CH 2; Ref: 18; 19; 20]

3.37 Silicon Ethoxide

Silicon Ethoxide is simply another name for TEOS [ethyl silicate]. It occurs with the dehydrogenation of Ethanol, this process also produces acetic acid which dissolves silica. The dissolved silica combines with the ethanol as metal oxides act as catalytic agents for the reaction. Remember that the rate of silica dissolution is accelerated by the anoxic conditions.

Another possibility that must be considered is the formation of Silicon Tetra acetate by a similar reaction, once again promoted by oxide catalysts. Silicon Tetra acetate degrades to TEOS and Ethyl acetate.

So it seems that man's synthetic method of making opal may not be that much different to how nature formed opal, if you only look at the organic end products that are required for opal formation.

As the organic matter breaks down it also releases more acetic acid, carbonic acid and ammonia (aq), and also some in the form of polyamines which are known to accelerate silica polymerization.

It is highly likely that in nature that many enzymes will play a large part in the decomposition of the organic matter to produce the organics for opal production.

However the main chemicals seem to be:

- Ammonia [NH_3, Urea, Polyamines]
- Acetic acid [dissolves silica]
- Formaldehyde
- Ethanol & Methanol [Methane]
- Metal Oxide catalysts [CH 4]

- Water for TEOS hydrolysis & condensation reactions
- Nuclearite [attracts the silica ions; bone, wood, shell (Bacteria?)]
- Magnesium thinning agent Silica.

Not to forget of course the opal clays and the Potassium feldspar as a source of the potassium ion which acts as an additional promoter for the hydrogenation reaction of the ethanol?

Opal does exist worldwide, but the question is where did the amino acids originate that caused the formation of opal. Noah's flood gives us the simple answer to this question. As all the plants and animals died under the flood waters they became buried by layers of sediment as the floodwaters began to recede. The sediment that covered this plant and animal matter would have been local sediment. This is why opal does not occur everywhere.

The plant and animal matter would not have been evenly distributed but would be deposited in large amounts in some areas and absent in other areas. Opal only forms where the sediment was opal dirt, in other words the right mixture of clays. It is the interaction of the Amino acids on the clays with moisture and an arid climate that produced temperatures around 40°C which helped to speed up the chemical reactions. Heat nearly always speeds up chemical reactions. So this has set a limit on the temperature of opal formation to under 100°C. The heat would be limited to climate, ground heat and heat of decomposition and possible a little exothermal heat. The other limiting factor is that above about 80°C the amino function would be impaired. The fact that opal forms underground indicates that the temperature was between 15 - 40°C and that light is not required for the reaction to take place. It is not a photochemical process.

Further evidence that amino acids are the electrolyte responsible for opal formation is the fact that amino acids have been found on fossils. They survive the decomposition process. This also suggests that the timeframe of fossils is thousands of years not millions. Amino acids would not have survived millions of years. We know that the time of Noah's flood was approx 5000 years ago.

There is also the recent formation of opal to consider. There are many stories of the recent formation of opal. These include fence posts not older than 5 years becoming opalized below ground level. The skeletons of peoples pets becoming opalized.

The Opal formation process will continue even today when the conditions are right. This alone puts to rest the old theories of opal formation. The real problem with the old theories is that they were never really theories in the first place. Real theories have to be proven to be accepted.

The Opal Miner's Cat

3.38 Example of Diagenesis

In 1896 an old opal miner's cat died so he buried the cat in the floor of his opal mine. The cat was placed in his hat and then buried at the bottom of his mine. A few days later the miner decided to retire from opal mining abandoning his mine. Some 50years later another man took over the opal mine. It was not long before he discovered the miner's cat. Only the bones of the cat remained but they had been turned to a pinkish coloured precious opal [Ref: 10, page 22].

How had the cat's bones become opal? The cat's body decomposed providing water and the decomposition product amino acid which transformed the opal dirt and the bones of the cat into precious opal. The carboxylic acids degrade to Carbonic acid which dissolves the silica and the ammonia produced catalyzes the formation of regular silica spheres which grow and then form a gel catalyzed by the $Al(OH)_3$ that has formed by hydration after being released from its silica bonds. This is the beginning of polymerization.

It is not only the decay process that is important but the conditions in which the decay takes place. The ideal conditions for opalization to occur would have been abundant in aluminosilicates, anoxic in nature, containing a rich organic source in a neutral pH environment. It is in these conditions that silica rapidly dissolves. Silica-organic-acid complexes promoted by these environments are responsible for this rapid dissolution of silica. This new information is in direct evidence to support the Noah flood idea for the formation of opal [Ref: 37].

These are the conditions we would expect of buried organic matter, the opal clays or aluminosilicates are essential as the source for the silica and aluminium. The fact that the pH is neutral could be somewhat of a novel idea, as silica dissolution has in the past been associated with either strongly acidic or strongly alkaline conditions.

3.39 Investigation of Decay End Products

In the Author's opinion based on the analysis of this organic chemistry info, we can expect that the main components of organic matter decomposition that influences opal formation to be the same as the end products of Organic matter decomposition; these three components are:

- CH_2O:
    - Glucose [Empirical Formula] (Ref: 32)
    - Carbohydrates
    - Acetic acid [molecular formula]
- Ammonia NH_3
    - From Amino acids NH_2 source
    - Could also be from Urea
- Orthophosphate (PO_4) phosphoric acid
    - Natural source bone

If the CH_2O or H_2CO is Formaldehyde, then it is providing a very strong dehydrating agent, one which is used in Taxidermy. Unlikely to be formaldehyde as it is a preserving agent.

Further supporting evidence of OM end products is provided

by the "Methane Catastrophe": It is interesting to note that Phosphoric acid is another Polymerizing catalyst. It is one of the most widely used Polymerizing catalysts in industry today. The other known fact about Phosphoric acid is that it causes fluorescence, possibly through Phosphorylation [See Ch 4. Opal Fluorescence].

"The equation employs a sort of "averaged" organic compound as a starting point. This compound employs the "Redfield ratio" of carbon to nitrogen to phosphorus -- 106 C: 16 N: 1 P -- that is typical of many organic compounds.

$C_3 106H_3 2630_3 110N_3 16P_3 1$ $+ 53 SO_3 4^2 + 14 H_3 20$ □
("averaged" organic compound) +   (sulfate)   + (water) (yields)

$53 H_3 2S$ $+ 106 HCO^3 + HPO_3 4^2 + 16 NH^4+ + 14 OH^$
(hydrogen sulfide) + (bicarbonate) + (phosphate) + (ammonium) + (hydroxyl)

(Kempe and Kazmierczak, 1996, Figure 4, p. 74, as modified from Kempe, 1990; the "phosphate" is monohydrogen orthophosphate.)"

[Ref: 33; Methane Catastrophe]

It is interesting that this particular scenario also provides us with the hydroxyl ion which is of utmost importance in dissolving silica.

There is a rather high concentration of Hydrogen sulfide, which can be very lethal, and has the smell of rotten eggs. If you intended to reproduce this reaction you would need to exercise extreme care.

At a pH above 9 the ammonium ion is converted to Ammonia gas.

Oxidation of fats [fatty acids] ; carbohydrates; and proteins results in the consumption of oxygen and releases carbon dioxide and water. The protein also releases Ammonia NH3 [Ref: 32, page 327].

Carbonic anhydrase can convert the CO_2 and water from fatty acid oxidation into carbonic acid.

In reference to the Methane Catastrophe there may well be a link to cellulose decomposition which can also generate Methane. [See previous chapter].

3.40 Biogenic Silica

There is a form of silica that has mostly been overlooked by researchers and scientists alike, that being biogenic silica from the decay of both plant and animals. Biogenic silica has very fine particle sizes. Plants and animals would not be able to absorb the silica otherwise. This silica is well placed, within the decay pool, to be converted to TEOS. The fact that only small amounts of TEOS are needed to catalyze the opal formation process lends considerable weight to this argument.

A general rule that applies to chemical dissolution is:

'The smaller the particle size, the easier it is to dissolve.'

It is similar to melting ice. A small piece of ice will melt much faster than a large chunk of ice, under the same conditions.

Biogenic Silica is more common than most people realize. It has been studied in Marine Sponges, Diatoms and some plants. The only reason that these organisms can take up silica is because of its very small particle size. [Refer to the above paragraph].

Most plants contain biogenic silica as catecholate complexes, these are actually silicon catecholate and can be described as silicon Alkoxides, which function as opal precursors. [See CH. 3, Catecholates].

The fact of the matter is that nearly every living thing requires biogenic silica to add strength to certain structures, such as, bones, shell, and wood.

The objection to TEOS being formed in nature has now been overturned, based on the available evidence. The existence of a TEOS catalyst Hydroxylase MMOHred in the organic pool indicates that TEOS formation in nature is no longer a possibility but a probability.

The possibility of TEOS forming from the decay of organic matter has been show to occur by at least two different routes. Let's take a look at both these methods:

- By the combination of dissolved biogenic silica and ethanol having $MMOH_{red}$ as the catalyst.
  [Metalloenzymes].
- Dehydrogenation of Ethanol in the presence of biogenic silica, forms silicon ethoxide, which is another name for TEOS.
  [Silicon Ethoxide].

3.41 Blood & Bone

In agriculture Blood & bone is used as a soil amendment; that is, it acts like a fertilizer. Blood & bone contains:

- Calcium Phosphate
- Calcium carbonate

It should be kept in mind that the bone is where red blood is manufactured.

Blood contains organic iron and has the highest concentration of Carbonic anhydrase in the body. Remember Carbonic anhydrase catalyses the formation of large quantities of Carbonic acid.

Bone also contains Gelatin which is a source of amino acids, which can be degraded to amines and carboxylic acids. Carboxylic acids degrade releasing H+ ions and CO_2. Carbonic anhydrase can efficiently convert these to carbonic acid.

Carbonate in the bone can also react with sulfuric acid to form Carbonic acid.

3.42 Polymorphism

The following items have been found opalized

- Bones
- Shells
- Wood

- Pine Cones

What do all these items have in common? They were once living or contained living organisms, and all contained Biogenic Silica. When these plants or animals died, they decomposed, leaving their amino acids behind to enter into chemical reactions. If this decomposing matter was buried in opal dirt they became candidates for carbonic acid ion exchange mechanism which dissolves surrounding silica and acts as an electrolyte allowing silica to penetrate the solid structures by ion exchange where it polymerizes to form opal gel.

Polymorphism of Glauberite and Alunite to opal are a result of these chemicals being washed out leaving a void which is later filled with dissolved silica carried by water, creating a cast, or jelly like mould [Elizabeth Smith; pg 48].

All of these items would be called Nuclearites, they attract the SiO_2 [opal gel] which exchanges ions with these materials, resulting in an exact copy of the original, then hardening to form opal. It is possible that Biogenic Silica is a nuclearite based on this information but yet another possibility exists, and, that is, bacteria a possible nucleating agent is probably inhabiting these organic structures. [See below].

3.43 Phosphorylation

This is a process that occurs in all living things, it happens at the amino acid level. Residues from this process may be partly responsible for opal fluorescence, however Sulfuric acid is also known to produce fluorescence as do some rare earth elements.

Other trace elements may also be responsible for opal fluorescence, such as thorium, uranium, Niobium or perhaps Hafnium.

3.44 Evolution of Strong Acid

The two main strong acids HCL and Sulfuric acids are evolved during the decay of plant and animal matter. HCL is the main stomach acid of most animals and is released during decay. Sulfuric acid is evolved with the breakdown of protein. These two acids on neutralization produce the salts of the acids, being Chloride salts and Sulfates. These two salts are able to decompose one another. This is a rather unique process often known as the "Double dissolution of salts".

These salts represent a commonality in the opal environment. Volcanic opal genesis occurs from such fluids. Sedimentary opal fields also have these same salts in common. More information on these salts will be found in the conclusion.

Organic Decay Diagram

The diagram below is used to illustrate the prolific nature in the evolution of some of the organic chemicals from the decay of plant and animal matter. This diagram is in no way complete but used to show the main end products especially the abundance of Carbonic acid, Ethanol, Ammonia, and Acetic acid in the decay process. Enzymes would also be present but are not shown in this diagram.

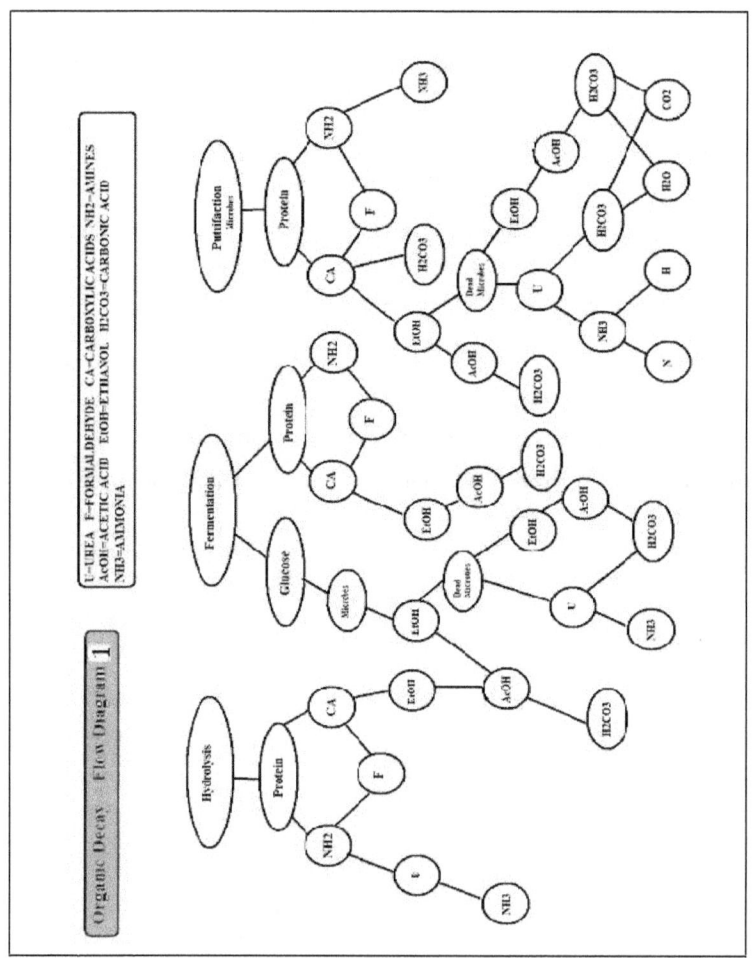

Humic & Fulvic Acids

3.45 Organic Decay Products

Humic and Fulvic acids are part of the humate content of almost all soils and as part of the decay process must be investigated to evaluate if they have any bearing on the formation of opal. Most people are aware of composting of organic scraps which produce the desired humic and Fulvic products, which can be added to garden soils to improve the quality of the soil. This decay process produces the compost, the dark moist remains of organic matter which becomes incorporated into the soil. Few people know much about the Humic and Fulvic acids which are the main components of the compost. This is towards the end of the decay process.

The purposes of our investigation is to decide if these acids are capable of dissolving clays, in order to establish weather or not they could be the elusive opalizing fluids.

> *"Humic acids (HAs) are naturally occurring biopolymers that are ubiquitous in the environment. They are most commonly found in the soil, drinking water, and a variety of plants. Pharmacological and therapeutic studies involving humic acids have been reported to some extent. However, when certain transition metals are bound to humic acids, e.g., iron and copper, they can be harmful to biological organisms. For this study, humic acids were extracted from German, Irish, and New Hampshire soils that were selectively chosen because of their rich abundance in humic material. Each sample was treated at room temperature with 0.1 M ferric and cupric solutions for 48 h. The amount of iron and copper adsorbed by humic acid was accurately*

quantitated using atomic absorption spectroscopy. The authors further demonstrate that these metal-loaded humic acids can produce deleterious oxidizing species such as the hydroxyl radical (HO*) through the metal-driven Fenton reaction. " (Ref: 43; Paciolla, MD & Jansen, SA).

It has been found that most humic complexes attach themselves to inorganic substances; such as clays and oxides with a smaller part being dissolved into solutions of the soil, essentially under alkaline conditions.

3.46 Humus

This term relates to the humic part of the humic substances that is not soluble in water at all.

Humin acid refers to the portion of humic substances that is not water soluble under acidic conditions below a pH 2 but become water soluble above this pH range.

3.47 Fulvic acid

Is the part of the humic substances that are soluble in water under all pH conditions. Fulvic acid is known to dissolve silica.

3.48 Phenolic acid

Is a part of the humic substances, and along with Fulvic acids have the ability to chelate and to bind heavy metal ions, reducing the toxic effect of these heavy metals.

The chemical characteristics of humic and Fulvic acids tend to vary greatly and thus do not have any set formula. The components of these acids will vary according to the organic matter content.

The origin of humic substances then results from the decaying vegetation

resident within the soils. Certain scientific studies have shown that it is the lignin content of plants that are proposed to be the source of humus.

It is more likely however that all organic building blocks of the vegetation play a role in humus formation. These include proteins, fats waxes, carbohydrates and resins.

Vegetation alone does not make up all the organic components of humic and Fulvic acids as decayed animal matter must also contribute its share to the humic substances pool.

Humic substances can have widely varying properties which are related to the functional groups located on the carbon chain. These could be:

- acidic (e.g. carboxylic acid and phenol),
- neutral groups (alcohol, aldehyde, ketone, ether, ester and amide).
- alkaline (e.g. amine, imine) or

Fulvic acids have some interesting properties like the humic substances have the ability to chelate metals and to act as ion exchangers, properties required to be an opalizing fluid.

Fulvic and humic acids are generally thought of as two entirely different substances with differing characteristics and properties. Fulvic acids tend to contain much more oxygen but much less carbon than the humic acids.

Typically Fulvic acids contain the following functional groups:

- *carboxyls,*
- *hydroxyls,*
- *carbonyls,*
- *phenols,*
- *quinones and semiquinones.* "(ref: 45).

3.49 Evaluation of Humates

The development of Humic and Fulvic acids are indeed part of the decomposition process but however seems that they are improperly digested. It is their ability to chelate minerals and heavy metals that is of interest. However, If these substances were properly digested, they may provide more consistent properties. Consisting of mostly carbon and hydrogen, it is conceivable that they may be the forerunners of crude oil production, simply requiring further digestion for the proper chemical reactions to occur, to fully dissolve the carbon and hydrogen molecules.

When fence posts become opalized we need to look at the soils in which this happens. The only evidence we are given is that the soils at the bottom of the post holes were dark soils containing a lot of organic matter. Organic matter tends to decompose to Humic and Fulvic acids. These acids may be involved in some types of natural opal formation. Fulvic acid in particular with its ability to dissolve silica and the fact that it has a very high Cation Exchange Capacity [CEC 1400], it is a highly effective ion exchanger. [Eckert, 22].

3.50 Summary of Humates

The main end products of decay are:

- Amines [Glutamine] – Ammonia NH3
- [CH_2O] Acetic Acid; or Glucose
- Bone → Phosphoric acid or orthophosphate [H_3PO_4]
- Carboxylic acids → Acetic Acid [CH_3COOH]

Although most of these will resolve into *Urea, Water and Carbonic Acid* according to Thomas Huxley.

3.51 Analysis of Humates

Nature it seems is quite capable of producing all the necessary ingredients and conditions for the formation of opal. It is clear that nature can dissolve silica to produce monosilicic acid along with the organics provided by the decay of organic matter, such as amino acids which interact with the SiO_2(aq) and water to form opal.

Something common to both crustacean shells and animal is Glucosamine which consists of Glucose & Glutamine. In marine life the main form is Glucosamine or Chitin [repeating units of Glucosamine]. In animals decay of amino acids produces something like 60% Glutamine and glucose. Glucosamine is contained in the cartilage of the joints and in bone marrow. It is very similar in structure to plant cellulose, which contains glucose & water.

Glucosamine seems to be very soluble in Acetic acid, but insoluble in water.

TABLE 10 – Common Plant & Animal organic chemicals

Both contain Amino Acids	(Carboxylic acids + Amines NH2)
Triglycerides	(Carboxylic Fatty acids + Glycerol)
Digestive Enzymes:	Protease breaks down protein (amino acids) Lipase breaks down fats (glycerol + Fatty acids) Amylase breaks down starch to simple sugars
Phospholipids & Phosphatids	Lecithin, is a surfactant and an emulsifier Phosphatidylserine
Sterols	Animal = Cholesterol Plant = Stigmasterol Vitamin = D3 (Cholecalciferol).
Amino sugars	Glucosamine
Glucosaminoglycans	Hyaluronic acid Dermatan Sulfate Chondroitin Sulfate Keratin Sulfate
Uronic Acids	Galacturonic acid; & Glucuronic acid (these are carboxylic acids).

Lecithin	Composed of Glycerol plus one or two fatty acid molecules and a phosphoric acid compound.
Biogenic Silica	Contained in Diatoms, the bones of animals and in the cellulose in plants, some plants also contain Silicon Catecholates.

The most abundant biopolymers in the world are:

- Cellulose: the most abundant.
- Chitin: The second most abundant.

Glucosamine is thought to breakdown in the presence of Silica & Potassium.

Related Topic Plant Decay

Wood chemistry; Cellulose; Lignins; Humic Acids; Fulvic Acid; Phenols; Petroleum Formation; Plant decomposition chemistry; wood decomposition; wood vinegar; chlorophyll; photochemistry; photolysis; wood Fenton reaction; silicification; Absorbents/Gelling agents; Amine Oxidase, ion-exchange; anion exchange, Silicification, silicified wood, Rapid silicification.

Related Topics Animal Decay

Taphonomy; Stages of Decay; Amino Alcohols [NH_2OH]; Autolysis; Paleontology; Earth Sciences; Geochemistry; Diagenesis

Diatom research -- Biosilicification:

Supporting evidence for Amino Acids in Opal genesis
[Amino Acids & ammonia in soils see: Analysis CH. 4]

a...*Biogenic Silica Patterning: Simple Chemistry or Subtle Biology*

By: Thibaud Coradin & Pascal Jean Lopez

Chembiochem Magazine 2003 Wiley

b...*Polycationic Peptides from Diatom Biosilica That Direct Silica Nanosphere Formation.*

By: Nils Kroger, Rainer Deutzmann, & Manfred Sumper

SCIENCE Magazine Vol 286 5/11/99

c…*Self Assemblage of Highly Phosphorylated Silaffins and Their Function in Biosilica Morphogenesis.*

By: Nils Kroger, Sonja Lorenz, Eike Brunner, & Manfred Sumper

d…*Silicification & Biosilification*

Part 6. Poly-L-Histidine Mediated Synthesis of Silica at Neutral pH

By: Siddharth V. Patwardhan and Stephen J. Clarson

Journal of Inorganic and Organometallic Polymers, Vol 13, No 1, March 2003.

e…*Biosilica formation in Diatoms: Characterization of native Silaffin-2 and its role in Silica Morphogenesis.*

By: Nicole Poulsen, Manfred Sumper and Nils Kroger

University Regenburg Germany

Related Websites Plant Decay

Petrification and alteration

www.ncsec.org/cadre2/team2_2/Articles/articles.htm

Silicification a type of petrification

encarta.msn.com/encyclopedia_761564197/Fossil.html

Further Reading Plant Decay

Fungal Decomposition of Wood, Its Biology and Ecology

By Rayner, A.D.M. and Boddy, L. (1988) Cost US$165

Fenton Reaction & water in decay process see Ch 2, 8, 10 (332-360), 11 (410-419).

Phil McCafferty, 'Instant petrified wood?',

Popular Science, Magazine, Issue October 1992, pp. 56-57

Further Reading Biogenic Silica

- Growing Opal Australian Style
  Snelling A.

  Creation (Magazine) 12(1):10-15, 1989

- Article: Rapid Opal Fits Bible Timescale

- Biosilica formation in diatoms: Characterization of native silaffin-2 and its role in silica morphogenesis.
  Nicole Poulsen, Manfred Sumper and Nils Kröger

  Lehrstuhl Biochemie I, Universitätsstrasse 31, Universität Regensburg, 93053 Regensburg, Germany
  http://www.pnas.org/cgi/content/full/100/21/12075

Exploring Bioinorganic Pattern Formation in Diatoms. A Story of Polarized Trafficking.

Chiara Zurzolo and Chris Bowler

http://www.plantphysiol.org/cgi/content/full/127/4/1339

- Silicatein: Cathepsin L-like protein in sponge biosilica. Katsuhiko Shimizu, Jennifer Cha, Galen D. Stucky, and Daniel E. Morse.

  http://www.pnas.org/cgi/content/full/95/11/6234

  Vol 95, Issue 11, 6234-6238, May 26, 1998

Related Topic Humic & Fulvic Acids

Fenton reaction; Polymorphism; Polymerization; Photocatalysis; Wood & Plant Chemistry; wood Fenton reaction; organic chemistry; organic oxidation reactions; Lignin Peroxidase (LiP); Gelation; Activators; Catalysts; Initiators; Gelling agents; Phosphoric Acid; Hydroxides; Hydroxyls; Radicals (hydroxyl, oxygen, sulfate); Electrolytes; Ionic polyelectrolytes; Hydrogen peroxide; Titanium dioxide; Thiosulfate; Iron sulfide; Acid sulfate; decomposition temperature; organic & inorganic chemistry; photolysis; photo disassociation; gelling ph.

Further Reading Humic & Fulvic Acids

Fungal Decomposition of Wood, Its Biology and Ecology

By Rayner, A.D.M. and Boddy, L. (1988) Cost US$165

Fenton Reaction & water in decay process see Ch 2, 8, 10 (332-

360), 11 (410-419).

Related Websites Humic & Fulvic Acids

PROSPER SUPERIOR HUME contains a wide variety of major and trace minerals, 21 amino acids and has a pH of 7.0 minimum. The Cation Exchange Capacity varies from 200 to 500 MEQ per 100 grams at a pH of 7.0. PROSPER SUPERIOR HUME is high quality humic acid, which has been extracted from dry humates

Websource: Circle One ® Online (Ref 6:42).

Bibliography: Chapter3

Plant Decay:

1. *Silicification and the conversion of Sinter to Chert*

Websource: http://www.abdn.ac.uk/rhynie/sinter.htm

2. *ESPM 134 Structure of Wood in relation to Decay*

Websource:
Http://mollie.berkeley.edu/~bruns/espm134/wood_decay.P
DF

3. *2-Chloro-1,4-Dimethoxybenzene as a Novel Catalytic Cofactor*

for Oxidation of Anisyl Alcohol by Lignin Peroxidase

By Pauline J.M. Teunissen & Jim A. Field

Department of Food Technology and Nutrition Sciences, Division of Microbiology, Wageningen Agricultural University, 6700 EV Wageningen, The Netherlands, 18th Dec 1997.

4. Title: *Copper Amine Oxidase*

Description: Website Article

http://pfam.wustl.edu/cgi-
bin/getdesc?name=Cu_amine_oxidN2

5. *Konjac Glucomannan* (KGM). (Info source 1:
http://www.glucomannan.com/glucomannan.htm

6. *Petrified Wood*

http://www.ncsec.org/cadre2/team2_2/Lessons/howDoesW
oodPetrify.

htm

7. *The Vasa's New Battle*

Acid in Wood

Magnus Sandstrom

http://www.fos.su.se/~magnuss/conclusions.html

8. *Biogenic Silica Patterning: Simple Chemistry or Subtle Biology?*

MIMIREVIEWS

P.J. Lopez & T. Coradin

ChemBioChem 2003, 3, 1-9

Wiley – VCH

9. *Sedimentary Geology: Rapid Wood Silicification in Hot Spring Water: an Explanation of Silicification of Wood During Earths History*

Hisatada Akahane; Takeshi Furuno; Hiroshi Miyajima;
Toshiyuki Yoshikawa and Shigeru Yamamoto
Publisher: Science Direct
Http://www.sciencedirect.com   6/06/08

Animal Decay:

10. *The World of Opals*   SL Lib Ref. 553.873 E19

Author: Allen W. Eckert

Publisher: Wiley 1997

11. *Modern Chemistry*   SL Lib Ref. S540.2

By Metcalfe; Williams; and Castka

Publisher: Holt, Reinhart, and Winston Inc., NY 1966.

12. Article: *Secrets of Growing Opal*

13. Article: *Some Aspects of Precious Opal Synthesis*

Publisher: Scientific Centre for Applied Research, Dubna JINR Russia

14. *Potch Opal & Its Genetic Significance*

Authors: V.V. Serdobintseva; D.V. Kalinin

15. *Introduction to Modern Colloidal Science*

Author: Robert J., Hunter.

Publisher: Oxford University Press, 1993

Lib Ref No. SL541.345

16. *Species-specific polyamines from diatoms control silica morphology*

Nils Kröger, Rainer Deutzmann, Christian Bergsdorf, and Manfred Sumper

http://www.pnas.org/cgi/content/full/97/26/14133

17. *Orthosilcates or Island Silicates*

Subject: Geology

http://socrates.berkeley.edu/~eps2/wisc/geo360/orthosil.html

18. *Mechanism of Growth of Supramolecular Crystals in Concentrated Suspensions of Mondispersed Spherical Silica Particles (MSSP)*

A.F., Danilyuk; V.V., Serdobintseva; D.V., Kalinin

International conference of Silica Science & Technology SILICA-1998, Mulhouse, France, Sept. 1998.

19. *McGraw-Hill Ency. Science & Technology* 8th Edition

Subjects: Silicon Vol. 16 pgs 452-460

Carboxylic Acids Vol 3 Pgs 252 - 255

Esters/Esterification Vol 6, Pgs 543- 545

Hydroxylase Vol ? Pgs.....-Diiron??

20. *Infracrystallization in monodisperse system of amorphous silica as a model of infracrystallization and growth of "photonic crystals".*

By: V.V. Serdobintseva; ICCG-13 Quoto Japan, June 2001.

21. *The Chemistry of Silica, (Solubility, polymerization, colloid...*

Published by Wiley, New York c1979

Author: Iler, Ralph., K

SL Lib Ref. 546.6832 I27

22. *Microbiology 315 Laboratory Manual*

Exercise #20: Preparation of Sauerkraut, Yogurt and Beer.

Http://www.towson.edu/~gekpenyo/315lab20.htm

23.. *Fats, Oils, Fatty Acids, Triglycerides - Chemical Structure*

Carbohydrates - Chemical Structure

Amino Acids, Peptides and Proteins - Chemical Structure

Antonio Zamora

ScientificPsychic.Com

24. *Making Gemstones Out of Household Kitchen Products*

Robert James FGA, GG

http://www.yourgemnologist.com/Kitchen/kitchen.html

25. *Think Glycolic, Properties,Uses Storage and Handling*

DuPont Specialty Chemicals, Belle WV

Http://www.dupont.com.glycolicacid

26. *The Missing Organic Molecules on Mars*

Subjects: Kerogen (Derived from Coal)

Amino Acids & hydroxyacids (Ruff Deg…).

Http://www.pubmedcentral.nih.gov/articlerender.fcgi?….

27. *Lysosomes*

Http://www.mnstate.edu/shima/lectures/cell1.htm

28. *Lysosomes in Normal cells & the Disease Process*

[Pages 1-8]

http://www.erin.utoronto.ca/~w3bio315/lysosomes.htm

29. *Lysosomes* [one page]

Disposal and recycling Units of the Cell

http://sln.fi.edu/qa97/biology/cells/cell7.html

30. *Amine Oxidases*

http://pfam.wustl.edu/cgi-
bin/getdesc?name=Cu_amine_oxidN2

31. *Decomposition of Organic matter* (Pgs 1-2)

http://www.ozestuaries.org/indicators/Def_decomposition.ht
ml

32. *Elements of Organic Chemistry* (Pages 81; 300 & 327)

International Student Edition 1967

John H. Richards; Donald J. Cram; George S. Hammond

McGraw-Hill Kogakusha, Ltd, TOKYO.

33. *Methane Catastrophe*

http://members.dcn.org/dorritie/methane%20catastrophe.ht
ml

34. *Biotechnology Opens New Routes To Nanofabrication of Silica,
Silicones, Metallooxanes and Organometallics*

Website Article By David Kisailus

Materials Research Lab, California NanoSystems Institute,
University of California, Santa Barbara, CA 93106

35. Thomas Huxley's Autobiography and Selected Essays

Thomas Huxley

www.alcdsb.on.ca/literature/THOMAS,
HUXLEY/AUTOBIOGRAPHYANDSELECTEDESSAYS.
HTML

36. *Clay Sedimentology*

Author: Herve Chamley

Publisher: Springer Verlag NY 1989

37. *Fate of Silicate Materials in a Peat Bog*

http://www.osti.gov/energycitations/product.biblio.jsp_id=55
39161

38. *Substrate Hydroxylation in Methane Monooxygenase Quantitive
Modelling via Mixed Quantum Mechanics/Molecular Mechanics
Techniques.*

By: Benjamin F. Gherman., Stephen J. Lippard., Richard A.
Friesner.

Vol 127 No. 3, 2005 pg 1025

Journal American Chemical Society [J.A.C.S.] Articles
29/12/2004

39. *Modelling non-heme iron Proteins*

By: Chuan He., and Yukiko Mishina

Current Opinion In Chemical Biology 2004

www.sciencedirect.com

40. *"The interaction of Ferric Ions with Silicic Acid."*

Hazel F., Shock, R. U. Jr., and Gordon, M.

American Chemical Society Journal., 71 2256-2257, 1949

41. *"The Encyclopedia Brittanica"*

Vol 23 pages 677:2a
Subject: Dissociation of Ammonia with Iron Clusters

Humic & Fulvic Acids:

42. *New Applications Of Redox Reactions For In-Situ Groundwater Remediation.*

By: David B. Vance
Websource: http://www.2the4.net/ThreeRedox.htm

43. Abstract: *Generation of hydroxyl radicals from metal-loaded humic acids*

By: Paciolla, M.D.; Jansen, S.A.[Temple Univ., Philadelphia, PA (United States). Dept. of Chemistry]; Davies, G.[Northeastern Univ., Boston, MA (United States). Dept. of

Chemistry]

Subjects: 54 Environmental Sciences ; Soils; Humic Acids; Iron; Copper; Soil Chemistry; Hydroxyl Radicals; Land Pollution

Websource:
http://www.osti.gov/energycitations/product.biblio.jsp?osti_id=355534

44. *Some Humifulvate ® Science*

© 2003 - Fulcrum Health Limited

Websource: http://www.fulcrumhealth.co.uk/page%2016.htm

45. *Circle One ® Online*

Subject: Humic Acid
Websource: http://www.circle-one.com/superiorhume.html

4 CHEMICAL REACTIONS

Fermentation Reactions

4.0 General Information

Fermentation and esterification reactions tend involve alcohols and some of their reactions. What then is an alcohol? Simply stated it is a pure carbohydrate.

4.1 Fermentation & Yeasts

Fermentation is a chemical reaction catalyzed by enzymes; these enzymes are produced by yeasts. It is an Anaerobic process (without oxygen) closely related to decomposition and composting. The end product of these processes is humus.

If fermentation is the answer to producing opal and the previous chapter strongly suggests that it is; then, it is important to understand yeast a little more, as yeast produces the catalyst for fermentation.

All sugars can be fermented but the fastest fermenting sugars take approx 48 hours, 2 days, these sugars are:

- Glucose
- Sucrose (table sugar)
- Molasses

Other sugars can also be fermented but can take up to 7 days for the fermentation process to be completed.

4.2 Fermentable Media

Apart from sugars other media that can be fermented are:

- Carbohydrates = Glucose
- Proteolysis: Protein Decomposition/Amino acids (Meat).
  Ammonia rather than ethanol is the main end product.

  Carboxylic acids may also be produced.

These media are available from both plants and animal decomposition.

4.3 The Two Types of Fermentation

- Aerobic [occurs where oxygen is available]
- Anaerobic [occurs where oxygen is depleted]

Typical Aerobic fermentation processes include our modern beer brewing and alcohol production methods, which use sugar glucose or carbohydrates as the fermentation media.

Composting is also an aerobic process which attracts aerobic bugs and microbes.

Anaerobic fermentation can occur where the oxygen is depleted [anoxic]; an example of such an environment would be burial of organic matter. This burial may occur in wet areas such as swamps and marshes or in dry soil conditions both of which will be without oxygen. In such an environment anaerobic microbes thrive, some of these microbes cause fermentation, however the products of anaerobic fermentation can be different to the products of aerobic fermentation.

Sugar fermentation produces mostly ethanol however protein fermentation produces a number of products, including ammonia and carboxylic acids, which are known to be involved in opal formation.

4.4 The Origins of Yeast

Yeast is isolated from certain fungi but it may also be produced by some decomposing microbial bacteria; fermentative microbes. The fungi that produce yeast are however plentiful and exist almost everywhere. These fungi feed almost exclusively on sugar.

4.5 Products of Fermentation

In commercial fermentation, it is interesting to note that Anaerobic Fermentation tends to produce a mixture of organic products. These include organic acids which lower pH, such as:

- Formic

- Propionic
- Acetic and
- Lactic

A variety of amines are also produced which tend to raise the pH, as well as Esters and Ethanol. Remember that carboxylic acids such as Acetic acid plus amines gives ammonium salts, which are catalysts in opal gel polymerization.

Ammonia and Amines are dehydration agents.

Yeasts themselves also produce amino acids which when fermented produce ammonia through a process known as proteolysis. If extra ammonia was needed it would be possible to add protein to the mix.

Proteolytic (digestive) enzymes break peptide bonds.

Amines & polyamines produced by diatoms and some sponges are implicated in the biogenic formation of Opal.

4.6 Main ingredients for Opal formation

- TEOS
- Ethanol [ethyl alcohol] plant derived (organic)
- Ammonia [catalyst], or Carbamate (UREA).
    Ammonium salts.
- Water.

It is probable that fermentation can provide all these ingredients. TEOS is produced by the combining of Ethanol with Silica. The main ingredient of fermentation is ethanol when combined with opal dirt (clay), it should result in temporary TEOS production. This would hydrolyze quickly because of the presence of water.

4.7 Relevance to Opal formation

Brewing is using the fermentation process to produce alcohols which consist of varying degrees of ethanol. Brewers often control the pH of the brewing process by using additives called Brewing salts. The five main brewing salts are:

- Gypsum $CaSO_4$ or Calcium Sulfate
- Calcium Carbonate $CaCO_3$ or Chalk
- Potassium Chloride KCL
- Table Salt NaCL
- Epsom Salt $MgSO_4$

It is probably no coincidence that in the opal fields the same electrolytes are found in the concentration of 3% of solution. This is the same concentration in which they are used in brewing. This indicates a possible fermentation process. TEOS contains both Si and Ethanol, it is probable that the fermentation process binds them together. TEOS begins the opal polymerization process and sustains polymerization until dehydration polymerization can complete the polymerization process. Monosilicic acid is probably not reactive enough to maintain the polymerization process.

Keep in mind that chlorides accelerate enzyme activity. Other fermentation processes use different chlorides.

In the production of Sauerkraut to provide optimum conditions for fermentation Sodium Chloride (table salt) is added at 3% concentration.

The temperature range for fermentation also falls within the range for that of opal formation (15-45).

4.8 Yeast activators/ accelerators

- Brewing salts accelerate yeast activity
- Chlorides accelerate enzyme activity in the yeast, these are often incorporated in brewing salts
- Temperature, ideal temperature range for yeast activity is between 15-45C. Below 15C the yeast becomes inactive, dormant. If the temperature is much above 45C it will kill the yeast. *Recommended temperature range is normally 15-30C.*
- Potassium from K-feldspar may also accelerate yeast action.

4.9 Two Types of Yeast considered

Two types of yeast will be considered because Turbo yeast has become the dominant yeast used in the alcohol industry, because it is supposed to have greater alcohol yield compared to brewers yeast. However Turbo yeast contains nutrients (yeast accelerants) but these may not be suitable for production of opal.

The opal environment however contains 3% chlorides and sulfates which just happen to be BREWING SALTS and just happen to be used in the same concentration; 3%; as found in the opal grounds.

It is then important to trial Brewing salts which means using plain yeast for brewing such as brewers yeast, otherwise there may be too much nutrient in the mix which could hamper the fermentation process.

4.10 Sterilization

When using yeast or fermenting you should always sterilize all your containers and anything that will come into contact with your ingredients.

Sterilizing helps to prevent the growth of unwanted microbes.

This can be done with boiling water or by a sterilizing agent such as Sodium Bisulfate which you will find in the brewing section.

4.11 Analysis of Fermentation in Opal Formation

Fermentation is definitely a part of the decay process and depending on the conditions can sometimes play a larger than usual role by fermenting not just the sugars, but also the protein, in which case more ammonia would be produced than normally expected. Yet it appears to me that the role of fermentation is limited to breaking down organic matter to provide raw materials for the opal process.

Further research has uncovered the fact that TEOS may not formed by the fermentation process because there is too much water, water inhibits the formation of esters, by most processes, unless there is a very strong dehydration process involved or if the water is bonded making it unavailable for reaction. TEOS is a silicon ethyl ester. See chapter on Esterification.

Scientists and geologists remain adamant that TEOS does not form in nature. This could be correct, but the possibility of TEOS forming in nature deserves further investigation. Highly probable is that monosilicic acid could be the precursor to opal formation, as shown above [Leo & Barghoorn] or form TEOS by chemical reaction [See: Silicon Ethoxide].

The Esterification reaction will also be investigated as a possible pathway to the formation of TEOS.

Esterification Reactions

4.12 General Information

The process of esterification is the reaction between an acid and an alcohol and is normally catalyzed by a mineral acid, the most commonly used is Sulfuric acid. The most commonly used reactant acids are the carboxylic acids (Acetic acid = CH_3COOH). One of the most common alcohols is Ethanol (C_2H_5OH). [For other alcohols see Alcohol Types,].

Acid + alcohol → ester + water

Silicic acid + Ethanol → TEOS + water

H4SiO4   + C2H5OH → Si(OC2H5) 4 + H2O

This reaction can be speeded up using a few drops *sulfuric acid catalyst*, and gentle warming, do not overheat.

There many types of esters that can be made depending on the type of acid & the type of alcohol used. Here are a few:

TABLE 11 -- Esterification		
Alcohol	Acid Type	Ester
Ethanol	Ethanoic	Ethyl Ethanoate
(Ethanol)	*(Acetic)*	*(Ethyl Ester)*
Methanol	Butanoic	Methyl Butanoate
Methanol	Salicylic	Methyl Salicylate
Propanol	Ethanoic	Propyl ethanoate
Octanol	Ethanoic	Octyl ethanoate
Ethanol	Butanoic	Ethyl butanoate

Concentrated Acetic acid (ethanoic acid) is used as a wart killer.

Alcohol Concentrations

Absolute Alcohol = contains less than 1% water (ANHYDROUS)

200 Proof = 100% alcohol

192 proof = 96% alcohol

190 Proof = 95% alcohol

Pure alcohol does not mean ANHYDROUS, free of other impurities.

Ethyl Alcohol is pure alcohol, not necessarily ANHYDROUS

Vodka is not Absolute alcohol

Esters can be made by:

- An acid & an alcohol
- An acid anhydride & an alcohol
- An acid chloride & an alcohol
- An acid & unsaturated hydrocarbon (olefin; acetylene)
- An Ester & an alcohol
- Two different esters

4.13 TEOS formation by Esterification

Finding the secrets of TEOS formation have not come easily with endless hours of research ending fruitlessly. It was not until I changed my focus to the TEOS formula that a plausible explanation became apparent. The TEOS formula is $Si(OC_2H_5)4$; taking the bracketed part of the formula OC_2H_5 and searching on the internet I discovered that this part of the formula is called Ethyl Ester or Ethoxide. In order to make TEOS you need to be able to make Silicon Ethyl Ester. This requires the Ethanol to be water free, that is ANHYDROUS Ethanol, or absolute alcohol. There are a number of ways of dehydrating Ethanol, the first is to use a zeolite of 3A size. You can also use Glycerin or Calcium hydroxide, but these are heat treatments similar to distillation in heat requirements; not possible in the natural environment.

The Ethyl part of the formula is represented by the Ethanol. Carboxylic acids + alcohols form esters. Therefore Acetic acid (vinegar) plus Glycerin should form an ester. This ester plus the anhydrous alcohol should then form Ethyl Ester. Anhydrous Ethanol plus acetic acid forms an Ethyl Ester.

This Ethyl Ester then needs to be reacted with Silica to form the TEOS.

Once TEOS forms it then needs to be hydrolyzed, however the esterification reaction also produces water, this should be enough for the hydrolysis reaction to occur. This reaction begins the silica polymerization that leads to opal formation.

It may even be possible to convert Ethyl Acetate $CH_3COOC_2H_5$ to TEOS by splitting off the Ethyl Ester, and combining it with Silica.

CH_3CO = Acetate? [Acetyl Group](Cellulose + Acetic anhydride).

OC_2H_5 = Ethyl Ester (Ethanoate/Ethoxide)

4.14 How does ANHYDROUS Ethanol form in Opal environment

The fermentation process produces an alcohol that is 95% concentration at best with the other 5% being water. TEOS will not form unless the Ethanol is anhydrous, containing less than 1% water but in reality commercial anhydrous ethanol is only 98% alcohol. This poses the question of "how does anhydrous ethanol form in nature. According to Atkinson dehydrated ethanol has the formula $C_2H_5OC_2H_5$ and is described as isomeric ether. Distillation is one of the main processes used by man to reduce the water content of alcohol, however this does not occur in the natural environment.

There is a clue in a statement quoted by Iler, that the Australian opal fields are underlined by Bentonite beds, which prevent the opal from drying out. It seems to me that the Bentonite serves another purpose also, that being as an absorbent of water, more specifically it has soaked up the water from fermentation processes leaving anhydrous ethanol which combines with the acetic acid and silica from clays to form TEOS. It must be remembered however that the TEOS will undergo hydrolysis with the water produced from the esterification of the ethanol and acetic acid. It is due to the temporary nature of TEOS that the evidence of this process does not remain to be discovered.

Bentonite, especially the Sodium variety soaks up water, swelling as it absorbs the water. It does not absorb ethanol, and evidence of this fact is that it is used as a clarifying agent in the wine industry.

4.15 Peracetic acid degrades to acetic acid [vinegar] which is also called Ethanoic acid (water free) or glacial ethanoic acid (99.5% acetic acid), or even Methanecarboxylic acid. It is used to make ethyl ester, more specifically Ethyl Ethanoate.

4.16 Oxidizing Ethanol

Ethanol can be oxidized to acetic acid by using KMNO4 as an oxidizer. This sours the Ethanol, thus causing the formation of vinegar, acetic acid.

Ethanol + Potassium Permanganate → Acetic Acid + Water

The alcohol is first oxidized to an aldehyde, using a strong alkaline oxidizing agent, if the reaction is allowed to continue it will form a carboxylic acid. The above reaction formula is an example of this type of reaction.

Water tends to inhibit the esterification reaction. What then is acceptable water content for the esterification reaction to proceed smoothly, and unhindered by the water? If the water is absorbed by clay, however, it is possible that esterification may proceed.

4.17 Transesterification

The esterification reaction of fats and oils with an alcohol (ethanol) is called transesterification, it is probably named after Trans fatty acids.

This process is used in the production of Biodiesel.

The main source of methanol is probably from plant, decaying wood is a source of methanol. The Methanol and oil are converted to ethanol, and the reaction proceeds as an esterification reaction, mostly used to produce an ethyl ester. In nature there may be both esterification and transesterification reactions producing ethyl esters, and silicon ethyl esters [TEOS] if in contact with silica sources, such as clays.

4.18 Phenol

Phenols are an organic compound found in plants, and extracted from coal tar. They are used in the production of plastics, synthetic detergents, wood preservatives, drugs fungicides, petrol additives, dyes and explosives. Phenol can be estrified using anhydrous acids, acid chlorides, or acetic anhydride (phenyl acetate).

Phenols are strong decomposing agents.

4.19 Dehydration Processes

Dehydration is simply the process of drying out, and it can be achieved through natural processes such as evaporation or by chemical processes.

Some of the dehydration and chemicals that dehydrate are:

- Evaporation (aided by salts - arid temperatures)
- Adsorbents/Absorbents, limits avail water
- Sulfuric Acid
- Saponification &
- Emulsification by making the water unavailable to hinder esterification.

Alternative is to use a waterless Alcohol such as GLYCEROL or possibly Sugar Alcohols + Triglycerides (NO WATER).

4.20 Iron in Opal Genesis

As you investigate the formation of opal you cannot escape the fact that the opal environment contains a lot of Iron. If Opal formation is not caused by the Fenton reaction, then what role does Iron play in its formation. In my research I have noticed that there are many different types of opal but that almost all of them are associated with some type iron. This is a very strong indicator that Iron plays a part in opal formation.

One of the most reactive types of iron in the wild is Iron Pyrite. The significance of Iron Pyrite is that with hydrolysis it produces a lot of sulfuric acid.

Sulfuric acid is a very strong desiccating agent; that is a drying agent. This is of great importance for dehydrating the ethanol formed by fermentation processes; in order to initiate the esterification reaction.

Water tends to inhibit the esterification reaction but the presence of sulfuric acid gives credibility to my notion that esterification is the beginning of the opal formation process.

The Iron in the opal environment would have originally have been Iron Pyrite (Iron Sulfide $FeS2$). Through hydrolysis, the leaching of acid from the iron, other iron compounds would then have formed.

The Iron Sulfide is the originating source of all iron, the source of iron is organic, as previously discussed (see The Fenton Reaction).

It must be kept in mind that two chemical dehydration processes here are working in tandem, the absorption of water by the clays and possibly even by some of the shale, along with the sulfuric acid dehydration of the ethanol.

How does Sulfuric Acid dehydrate Ethanol? Sulfuric acid combines with the water in a process known as dilution. The water is absorbed into the acid, acting much like a solvent, thinning out the acid. The chemical formula remains unchanged, but there is now no longer any free water as it is now part of the acid.

The Sulfuric Acid required for the esterification reaction is concentrated Sulfuric Acid which is deemed to be 98% concentration and above. The amount used for esterification is very small, as it is only for catalyzing the reaction and for dehydrating the Mentholated Spirits so they become ANHYDROUS Ethanol.

It should not be necessary to add any Acetic Acid as some of the Ethanol will oxidize forming Acetic Acid. The Sulfuric acid in the mix should be sufficient to dehydrate the water formed by this reaction. FeS2 with water forms concentrated sulfuric acid plus the transition metal iron catalysts Fe2+ and Fe3+ as per the Fenton reaction.

Do not add too much Sulfuric acid to the mix as this could cause the ethanol to dehydrate to ethene (gas). The aim is to produce anhydrous ethanol for the esterification reaction which forms the TEOS within the mixture.

Esterification also produces water as a byproduct of the reaction which will cause hydrolysis of the TEOS. This will happen automatically. Once the mixture is capped, there should be no need to touch it for months by which time, the opal colours should be very obvious. You should see colour within two to three weeks. Once colour is achieved, your success should also be achieved, as long as you dry the opal properly (see: Ch. 4 Dehydration of Opal Gel).

However the role of Iron may extend much beyond what has already been stated above. It is the Iron Oxide that oxidizes the Al in order to form Aluminium Oxide which is pivotal to the formation of opal. It happens by the following formula:

$$Fe_2O_3 + Al \rightarrow Al_2O_3 + Fe$$

However high heat may be required for this reaction, without which it may still proceed but at a much slower pace. The oxidation of Al is discussed in the polymerization chapter.

If the Iron is combined with H_2O_2 it may also be involved in the breakdown of organic matter, i.e. protein [Fenton reaction].

4.21 Proteins & Amino Acids

The way to release Amino Acids from Protein is to use a mineral acid such as Sulfuric Acid. Amino Acids contain both carboxylic acids and amines. Sulfuric Acid releases the amino acids freeing them to enter into the esterification process with the available ethanol. The sulfuric acid also acts as a catalyst in the esterification reaction, possibly working in tandem with the Amines as catalysts. [Boiling maybe required].

Doc Brown describes Amino Acids as being Carboxylic acids (like ethanoic acid) but with one of the hydrogen atoms of the 2nd carbon atom replaced with an amino group (a nitrogen + two hydrogen gives $-NH_2$). Another hydrogen on the same 2nd carbon can be substituted with other groups of atoms (R) to give a variety of amino acids.

The simplest is aminoethanoic acid or 'Glycine'.

This indicates that Amino acids should readily enter into the esterification reaction with anhydrous ethanol to form esters.

Amino Acids may in fact be essential for the formation of

natural opal.

4.22 Esterification without DEHYDRATION

Alcohol + Carboxlic Acid → H_2SO_4 → Ethyl Ester

Glycerol + A Triglyceride → H_2SO_4 → Ethyl Ester

Substitute Glycerol $C_3H_5(OH)_3$ for ethanol C_2H_5OH

Substitute Triglycerides for Acetic Acid CH_3COOH

4.23 Acetates ?

Cellulose Acetate + Silicic Acid → Metallic oxides → TEOS & Acetyl group [Dehydrogenation]

It appears very likely that the above reaction may well take place where wood decomposes with opal clays. Wood contains absorbents that may well absorb enough water for esterification to proceed.

Leo & Barghoorn equate wood decomposition with Ethanol. Ethanol enters into the esterification reaction, however wood can also decompose to Cellulose Acetate in the presence of Acetic acid. Ethanol degrades to Acetic acid with the presence of water, or an oxidizing agent.

It may be that this reaction leads to the release of the absorbing agents.

4.24 Triglycerides

Myristic $CH_3(CH_2)_{12}COOH$

Palmistic $CH_3 (CH_2)_{14}COOH$

Stearic (lard) $CH_3(CH_2)_{16}COOH$

Oleic $(CH_2)_7COOH$

Linoleic $(CH_2)_6COOH$

See Chapter DECOMPOSITION Heading Saponification

SOAP may even substitute for Carboxylic Acid

Glycerol is an oily alcohol. It contains little water. All organic oils will eventually decay, the end product of this decomposition is Glycerol.

Another common name for Glycerol is Glycerin or Glycerine.

Drying oils belong to the triglyceride group.

4.25 Polymerization Types

- Condensation
- Dehydration
- Esterification

4.26 Processes of Opal birth

Burial in Opal Dirt

Autolysis [Lysosomes] -- body starts to decompose itself

Fermentation or Putrification of Organic matter

Hydrolysis & dissolution of Silica

Dehydrogenation of Ethanol by metallic oxides forms TEOS

Condensation & Hydrolysis of TEOS

Opal Gel Formation & Ion Exchange

Alcohol Condensation polymerization

Water Condensation polymerization

Drying in clay [clay becomes the molds]

Note: Wood Opalization; Lignin is basically alcohols, these only need to combine with a carboxylic acid such as acetic acid to form an Ester, the cellulose degrades to glucose which ferments to ethanol or methanol.

4.27 Chemicals like TEOS

TEOS $Si(OC_2H_5)_4$
Ethyl Acetate $CH_3CO\underline{OC_2H_5}$
Cellulose Acetate $CH_3CO\underline{OC_2H_5}$

Diethyl Carbonate $CO(\underline{OC_2H_5})$

Ortho-carbonic ester $C(\underline{OC_2H_5})_4$

Chemicals with a similar formula can often be converted to the target chemical by one means or another. It should be possible for at least some of these chemicals to be converted to TEOS. Man would not need to do this but it may happen in nature.

4.28 Esterification in opal forming

The fact is that most of the chemicals would be available for the esterification reaction to produce TEOS. The esterification reaction is severely inhibited by the presence of even small amounts of water. Water is required for the hydrolysis which results in the dissolution of silica. Almost all the theories of opal formation have water as an integral and indispensable part of the process. However water in the presence of FeS_2 produces concentrated sulfuric acid.

It seems that Science has dismissed the formation of TEOS in nature due to the fact that water is prevalent in nature and dehydration to the anhydrous state is very difficult. However the author still believes that TEOS does in fact form in nature, and most probably is the agent or precursor to opal in the natural environment. TEOS is an organic molecule and therefore it is more likely that it formed in nature long before man discovered how to synthesize it. Ethanol even in nature tends to form as 95% with the other 5% being water but it is possible that Carbonic acid may dehydrate this ethanol by splitting the water into H+ ions and OH- groups which react with the Silica, the alumina and other oxides present in the clays preventing water from reforming.

Thus the search begins to discover in nature the most applicable dehydrating agents and chemicals that will lead to TEOS being formed in the wild. My research into the natural formation of opal indicates that there are indeed natural absorbents and dehydrating agents which may work in tandem with solar evaporation, to bring about suitable conditions for the esterification reaction required for TEOS to be formed.

[See heading this Chapter: Dehydration Processes & Chemicals]

According to Leo & Barghoorn, another source for investigating is Monosilicic acid as a form of dissolved silica, the natural silica source for the opal formation process.

It is highly probable that the esterification reaction could be involved in the formation of opal. The Fenton reaction is also highly likely to be involved, producing not only the iron metal transition catalysts but also the highly reactive Hydroxyl ions, possibly even the sulfate radicals.

Ammonia Properties

4.29 General Information

In looking for the elusive opalizing fluid, it should be kept in mind that we are looking for something special. This has eluded scientists and geologists alike. No-one has found any electrolyte, or even traces of it in the opal environment. It must therefore be considered to be biodegradable. However we do have a start point.

It is clear from the Russian Stober method that the electrolyte for opal production is Ammonia [NH_3 or NH_4OH ion]. These ammonia species [concentration approx 29%] are commonly used in industry to produce silica spheres. Ammonia can be quite lethal at this concentration.

Ammonia is an organic chemical, is found in the natural decay of dead organic matter. Generally speaking Ammonia due to its reactivity, normally resolves quickly to forms salts and compounds with other chemicals, or is released as gas into the air.

Organic matter is a constituent of almost all soils and is even present in volcanic fluids. It is known to be involved in the opal formation process which will be discussed in more detail in this chapter.

This is not the electrolyte that Len Cram uses.

4.30 Ammonia Sources

The main source of Ammonia these days comes from the distillation of petroleum which is itself a product of the decomposition [decay]of organic matter, which is helped along by some of the following:

- the presence of: FeS_2 &
- Carbonic acid &
- Potassium; &
- Sulfates which release H_2SO_4.
- Iron species [Fenton reaction; breakdown organics]
- Clay catalyzed reactions [Smectite as catalyst; diagenesis= burial]
- Enzymes & Lysosomes [cell breakdown -- autolysis]

The organic matter would be amino acids, the main ones being:

Glucosamine (Glutamine + Glucose) which is a major common element in nearly all forms of opal polymorphs.

4.31 Sulfate & Chlorides

These two ammonium salts are found both in the volcanic and sedimentary environments.

Ammonium sulfate seems to be very slow to react possibly resulting in a weak electrolyte. Tends to form granules which are slow reacting.

Ammonium Chloride on the other hand should be a good electrolyte as Chlorides generally tend to form very strong electrolytes. It is normally in powder form and very soluble in water. At a 10% powder concentration it was used as snuff powder, or smelling salts.

4.32 Salts & Bases

Any Ammonium salts mixed with a strong base will release ammonia in solution. The Ammonium salts can be halides, sulfates, sulfides, nitrates, acetates or even phosphates can be mixed with strong bases such as hydroxides of sodium, potassium, lithium, or calcium, as well as calcium oxide or carbonate. A little heat would help the reaction along, producing ammonia gas in solution. There is plenty of Calcium Carbonate in the opal fields of Australia, with the two most prominent ammonium salts being Ammonium Sulfate and Ammonium chloride.

4.33 Amino acids are a source of both Carboxylic acids COOH & Amines NH_2. The Carboxylic acids break-down to Carbonic acid and the amines form ammonia, and ammonium salts.

4.34 Hydrolysis of Amino Acids involves breaking the peptide bonds through the addition of water, forming two molecules: a carboxylic acid, and an amine. This is the same as hydrolyzing simple amides, except that when dealing with proteins, the molecules carry a lot more baggage. In the lab hydrolysis requires high temperatures and either strong acid or strong base. In cells, this reaction takes place at body temperature and is catalyzed by an enzyme, since strong acid/base is not a practical consideration in living tissue.

Amino acids have been dealt with in detail the chapter on Organic Chemistry.

4.35 Urea

In mammals, urea is formed by the deamination of excess amino acids; a manganese-containing enzyme, known as arginase, is involved in this biochemical process.

Urea is a powerful protein denaturant. This property can be exploited to increase the solubility of some proteins. For this application it is used in concentrations up to 10M.

Also known as Carbamide or Carbamate has the formula:

$CO(NH_2)_2$ or CO_2NH_4; H_2NCONH_2 ; CH_4N_2O; $(NH_2)_2CO$; the ammonia is found to be in the form of an Amine NH_2, but more like a Carbamine:

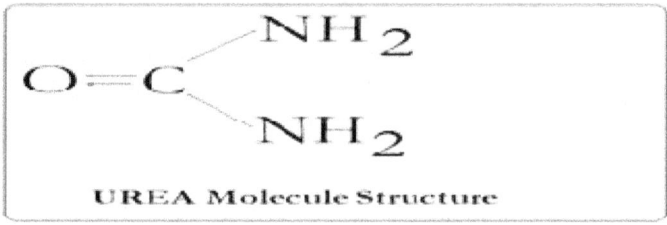

Figure 12. UREA molecular structure, clearly shows two Amine groups.

Contains both the necessary ingredients to transform clays to opal. The CO_2 which forms Carbonic acid in water and the Ammonia.

Urea dissolves in water at 40°C, it is an endothermic reaction, causing drop in solution temp, decomposes before boiling,

How Urea Converts to NH_4

There are a number of choices when choosing urea, there is the prilled form and the granular form. Urea is a white crystalline granular material containing 46% N. The granules are larger, harder, and less affected by moisture (high humidity) than the prilled form which used to be the primary form of urea. Consequently, granular urea has become a common form of urea for fertilizer blends.

Hydrolysis, the breakdown of urea in the presence of water, is a fundamental property of urea that greatly impacts the management of urea as fertilizer. The importance of urea and its hydrolysis as a source of organic chemicals which may impact on the formation of opal will be considered here and in the book conclusion.

Urea hydrolyses to ammonium carbonate very quickly when added to and incorporated into the soil according to the following reaction:

$$CO(NH_2)_2 + H_2O + urease \rightarrow (NH_4)_2CO_3\ 2NH_3 + 2H_2O + CO_2$$

This reaction formula however misses a vital stage of the overall reaction.

Before CO_2 & H_2O are produced there is the formation of H_2CO_3 or Carbonic acid. The formula should read:

$$CO(NH_2)_2 + H_2O + urease \rightarrow NH_3 + H_2CO_3 \rightarrow NH_3 + H_2O + CO_2$$

The Carbonic acid has a short life and quickly breakdown to $H_2O + CO_2$

And for this reason is often neglected in the reaction formulas however for investigating opal formation it could be essential. It is the NH_3 and the H_2CO_3 that are of the most significance as they are both implicated in breaking down clays. The pH is a prime factor in determining if the result is ammonium NH_4 or the ammonia ion NH_3.

Urease (a natural soil enzyme)

The ammonium carbonate is very unstable and decomposes to NH_3 + carbon dioxide. The enzyme, urease, greatly stimulates this reaction. Urease is a metalloenzyme of the nickel variety which catalyses the conversion of urea to ammonia and carbamate.

Urease [Urea amidohydrolase, EC 3.5.1.5] is found in:

- Plants
- Algae
- Yeasts
- Some fungi [filamentous]
- Soils
- Bacteria, and
- Microbes

The most common sources for industry are:

- Jackbean [Canavalia ensiformis]
- Soya bean [Glycine Max]

Urease is found in all soils in sufficient quantities to bring about rapid conversion of urea to ammonia (NH_3) which then in the presence of water converts to ammonium (NH_4^+). The ammonium then attaches to the negatively charged soil particles and behaves like any other ammonium-based source of nitrogen fertilizer. This whole hydrolysis reaction normally occurs within a 2 to 4-day period. If the urea is not incorporated into the soil, hydrolysis can also occur and NH_3 can then be lost to the atmosphere through volatilization.

4.36 Hydrolysis of UREA

The conversion of urea $(CO(NH_2)_2)$ to ammonium $(NH_4)^+$ occurs in a two step process. When the urea combines with water (hydrolyses) it forms ammonium carbonate $(NH_4)_2CO_3$. Ammonium carbonate is unstable and decomposes to form ammonia gas (NH_3) and carbon dioxide (CO_2). The ammonia gas produced is chemically identical to anhydrous ammonia. If the ammonia gas is in physical contact with water, it reacts to form the ammonium ion $(NH_4)^+$. If the ammonium ion is in contact with the soil, it is attracted to the clay and organic matter particles and is held in the cation exchange complex. http://lancaster.unl.edu/ag/factsheets/288.htm

Urea is also a common source of nitrogen. Highly soluble in water, urea hydrolyzes to carbonic acid and ammonia, given time. [Ref 8:9; Survey of Fertilizers].

4.37 Urea & Oxidizing Agents

Urea also decomposes in the presence of strong oxidizing agents, some of which are Fluorine, Nitrates, Potassium Permanganate, Hydrogen Peroxide, Sodium Hypochlorite [bleach], Sodium dichromate, Chlorine, Phosphorus Pentachloride and strong acids.

4.38 The Bicarbonate

This was once used as a baking powder but has fallen out of favour with other products now being used instead. Ammonium Bicarbonate easily breaks down in water at temperature range of 36 – 60C. On decomposition it produces Ammonia and carbon dioxide. More ammonia will be produced in alkaline solutions. More CO2 will be produced in acidic solutions. Superior choice to Urea in releasing ammonia.

Urea does not decompose as easily.

4.39 Ammonium Hydrolysis

In acidic aqueous solution ammonia NH3 gas is converted to the Ammonium NH4+ ion which often combines with other ions such as sulfates, changing states from gas to solid.

Theoretically, any Ammonium ion source can be used if it is reacted with the Hydroxide ion OH- as this result in Ammonia gas and water:

$$NH4+ \; + OH- \; \rightarrow \; NH3 \uparrow + H2O$$

A pH of 9 and above may be required.

[Modern Chemistry Ref: 1, 485]

Alkaline earth metals include: Beryllium (Be), Magnesium (Mg), Calcium (Ca), Strontium (Sr), Barium (Ba), and Radium (Ra). These Alkaline earth metals are very electropositive which makes them very good conductors of electricity. They are also very good reducing agents, the most abundant being Calcium & magnesium. Radium is radioactive and quite rare. Barium species are quite common in some hydrothermal waters.

When it comes to mixing your solutions it may be best to do two separate mixes; the clay feldspar and water; and the electrolyte mix. The electrolyte mix will contain the TEOS [soluble silica], Ethanol, ammonia and some Alkali metals; such as K or Na [as sulfates or carbonates] and Alkaline Earth metals such as Mg or Ca, can be in the form of carbonates, add these slowly, little by little as this will be an exothermic reaction. This will produce the electrolyte as mentioned in the shaded box. It is important that the pH of your final mix should be around 10 to keep the ammonia in the aqueous form as long as possible, to keep the electrolyte active. Above pH 9 the ammonium ion reverts to the ammonia(aq) form.

4.40 More Properties of Ammonia

Ammonia; especially liquid ammonia is a non aqueous ionizing solvent.

Evidence of this is Ammonia's ability to dissolve the Alkali Earth Metals including other electropositive metals such as Calcium, Strontium, Barium, Europium and Ytterbium. As it dissolves alkali metals it forms highly coloured, electrically conducting solutions [electrolytes] containing solvated electrons [Ref: 18]. Solvated electrons are free electrons surrounded by a cage of ammonium ions. This produces an electric pathway for Ion exchange to occur. Could this be THE ELECTROLYTE for opal formation and a strong reducing agent.

4.41 Solvated electrons [Ref: 21]

There are two chemicals that produce strongly solvated solutions. These chemicals are:

- Ammonia
- Lithium

In solutions (as in water) ions are frequently bound non-covalently with the molecules of solvent, and in that case are said to be solvated. An *ion* consists of one or more atoms and carries one unit charges of electricity.

Positively charged ions are called cations [hydrogen and metals] and those which are negatively charged are called anions [hydroxyl and acidic atoms or groups]. When a solution containing ions is made part of an electric circuit, the cations move toward the cathode, the anions toward the anode. This movement is called atom migration, which explains how ion exchange occurs. Atom migration happens in some chemical reactions without electrodes of any kind.

The presence of aqueous ammonia significantly increases the rate of silica dissolution by releasing Alkali Metals such as Potassium (K) and Sodium (Na) as free ions into solution at high pH, above 10. The K and Na ions are known to dissolve silica. The high pH itself is a factor in the dissolution of silica.

These conditions should result in very high percentages of the total silica being dissolved.

Ammonia directs the polymerization of colloidal silica to form opal gel not just another silica gel. Without ammonia, its salts or amines, opal simply would not form. Opal does not form everywhere even when opal clays are present, because ammonia was not available everywhere. Ammonia is the organic element in the opal formation process being derived from decayed or decaying organic matter whether it is plant or animal in origin. Protein reaction with Biogenic silica See: Figure 5 CH. 3 and [Ref: 24]

4.42 Formation of Al(OH)3 [Ref: 19]
Al^{3+} ion reaction Aqueous Ammonia:

Aluminum ion reacts with aqueous ammonia to produce a white gelatinous precipitate of $Al(OH)_3$:

$$Al^{3+} (aq) + 3NH_3(aq) + 3H_2O(aq) <==> Al(OH)_3(s) + 3NH_4+(aq)$$

[http://www.public.asu.edu/~jpbirk/qual/qualanal/aluminum
.html]

The significance of the formation of Al(OH)$_3$ has already been
mentioned in previous chapters, but this is how it forms under
the stober process of TEOS polycondensation.

If Ammonia dissolves Calcium from Calcium Carbonate, this
would leave the Carbonate ion in solution where it could
combine with available H+ ions to become Carbonic acid,
dissolving silica at alkaline pH, producing colloidal silica.
Ammonia's ability to dissolve alkaline metals and
electropositive metals means that the oxides or hydroxides
attached to these metal elements will be freed as radicals which
will help to dissolve more silica, could even act as the main
solvent for silica dissolution to feed the polymerization
process.

Extreme caution should be taken when using ammonia or its
associated products. Oxidation of ammonia results in
dinitrogen and reduction results in dihydrogen, these are slow
reactions. In the presence of alkali metals reducing conditions
will prevail, the result is the slow decomposition to the metal
amide and dihydrogen. Oxidation of liquid ammonia is
normally quite slow, there still exists a risk of explosion if
transition metal ions are present, as these may act as catalysts
[Ref: 18].

4.43 PolyAmines

In the natural opal environment ammonia may be present in
the form of Amines as a result of Amino acid decay. Amino
acids can take a very long time to breakdown, however amines

are available to us in the form of CATIONIC SURFACTANTS; these contain:

Primary Amines

Secondary Amines

Tertiary Amines &

Quaternary ammonium.

Cationic surfactants can be used alone or in addition with an inorganic or organic base to increase pH during the gelation or coagulation.

The preferred catalyst in industry however is water/Ammonia in a concentration of between 25 – 30%. The amines may be much slower to give the same result, as the amines may need to be converted to liquid ammonia in the water solution, and getting the right concentration could be much more difficult if you have no experience with these types of chemicals.

The case of the miner's cat is evidence that Amines should be a good substitute for liquid ammonia. It is possible that the amines will degrade to ammonia anyway. Amines and polyamines form from the decay of organic matter. However there is still much research to do relating to the reactions of amines and polyamines.

Recent research into certain marine sponges and diatoms has revealed proteins which form opal, these proteins are called silaffins. They exhibit a remarkable similarity to some well known amino acids such as:

- Poly-L-Histidine
- Poly-L-Serine

These organic amino acid residues are produced by the decay of organic matter, both plant and animal. In other words they are common elements of the decay process.

Amines may in fact be a very good gelling agent for silica polymerization. There are other agents involved in silica polymerization but few are directly linked to opal formation, apart from amines/ammonia and Aluminium hydroxide. We don't know how many years it took for the cat's bones to become opal, it would probably have been 1-2 years depending on drying time. In arid areas this would be expected to be a relatively short time, due to evaporation and dehydration by the clays.

Silica spheres form in the presence of:

- Ammonia NH_3/ Amines NH_2
- Ammonia salts
- Poly-L-histidine [Copper Amine Oxidases on amines]

Primary amines in the presence of CAO's in the Fenton reaction would add more H_2O_2 to the Fenton reaction and release NH_3 to initiate silica sphere formation along with the three Poly-L-histidine residues of the CAO's themselves.

4.44 Ammonia & Oxides [see Table 10]
See also Chapter on Wood Chemistry relating to Petrified wood and the chemicals required for petrification to take place. Compare these chemicals:

- The chemicals that ammonia dissolves.

- Petrification chemicals
- Volcanic Ash
- Wood ash
- Volcanic fluids

Mineral water

4.45 EDTA [Ref: 20]

Ethylene diamine tetra acetic acid [EDTA] has virtually replaced the combined Citric & Tartaric Acid complex once widely used in industry as ion exchange and chelating compound.

It is a strong complexing agent, reacting with many metallic ions to form soluble chelates. It is commonly used to keep alkaline earths and heavy metals in solution.

Chelates are metal ions in aqueous solution which are completely solvated or hydrated to give an aquo complex (see ligands) [3:540]

It can be prepared a number of ways from Ethylenediamine an amino acid. [1]Ethylenediamine plus formaldehyde & sodium cyanide in basic solution or from [2]Ethylenediamine & sodium chloroacetate. Below is shown the structure of EDTA.

Ethylenediamine is an amino acid and formaldehyde forms from amino decay, is sodium cyanide or sodium chloroacetate found in nature. Could EDTA be formed in nature.

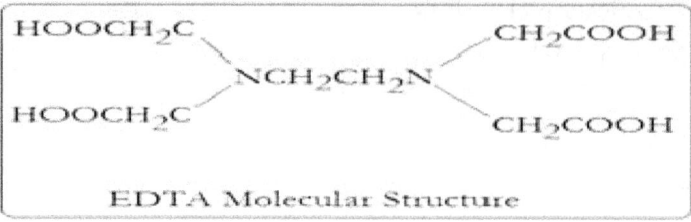

Figure 13. This diagram shows the Quaternary structure of the EDTA Molecule.

In order for EDTA to substitute for ammonia it must produce the same conditions creating an electrolyte with solvated electrons. We know that it creates the solvated electrons but do they function as an effective electrolyte and reducing agent as ammonia does.

A reason for considering EDTA apart from its electrolyte properties has to do with its chemical structure. It contains four Acetic acid molecules and an Amino acid, Ethylene diamine. The diamine may be a safe substitute to liquid ammonia which can be very dangerous to handle.

It is known that some amino acids both natural and synthetic have the ability to produce silica spheres from soluble silica. EDTA is very stable but to obtain maximum benefits in opal synthesis it must decompose releasing Acetic acid and the diamine. Acetic acid is known to be silica solvent whilst diamines can replace ammonia as the silica sphere growth catalyst.

One known method of degrading EDTA is by using ZEA.

This reaction apart from releasing the Diamine and the Acetic acid also produce a number of reactive oxygen species:

- O_2^-
- H_2O_2 and
- HO^-

Other possible ways of degrading EDTA could be by employing incompatible agents such as:

- Strong oxidizing agents
- Strong bases
- Copper/Copper alloys
- Aluminium [evolves Hydrogen].

Never use incompatible agents unless you know how the reaction will unfold. Many agents are listed as incompatible simply because they decompose the target chemical, however some decompose explosively or with lethal fumes.

There is also an ammonia version of EDTA.

4.46 Ammonium Thiosulfate $[(H_3N)_2H_2O_3S_2]$

This chemical is used as an electrolyte in the processing of film, it is used as a fixing agent by reaction with silver halide converting it to silver metal. It is also used as a metal cleaner.

Thiosulfates have been discovered in hydrothermal waters and also in petrochemical environments.

Also called; Thio-sul; Diammonium Thiosulfate; Ammonium Hyposulfite; Thiosulfuric acid diammonium salt.

4.47 Ammonium Persulfate $[(NH_4)_2S_2O_8$

Has high solubility in cold water and is a very powerful oxidizing agent often used as a radical initiator for polymerization process. It is also used to etch copper printed circuit boards.

Other names include; Ammonium Peroxydisulfate; Peroxydisulfuric Acid, Diammonium Salt; Diammonium peroxydisulfate

4.48 Ammonia Decay

It has recently been discovered that the ammonia molecule dissociates when it attaches to an iron cluster. These clusters are found as Fenton iron species and as biological clusters such as $MMOH_{red}$.

4.49 Ion exchangers

Whilst EDTA has proven to be a good ion exchanger there is an even better one. Sodium Gluconate $[C_6H_{11}O_7Na]$ is an excellent ion exchanger. It can be made by an inorganic reaction or by anaerobic fermentation as would happen in decaying buried matter. However, the opal ion exchanger is Montmorillonite clay.

4.50 Analysis of Ammonia Sources

As opal forms in clay soils then the ammonia sources must be soil related sources. The two main sources of ammonia in soils are Urea and Amino Acids from both plant and animal decay, the release of which is accelerated by microbes.

Nitrogen exists in the surface layers of almost all soils as decaying organic matter with the rest being in the form of the NH4- ion that is often held within the structure of clay minerals. Plant matter and other debris buried in the soil exist as Amino Acids. Typically cultivated soils contain between 0.06 and 0.3% Nitrogen but humic soils may contain as much as 3.5% Nitrogen.

Solvated electrons seem to be the Key for producing a superior electrolyte having a conductivity approaching that of the metals. These solvated electrons are produced in the presence of Ammonia or Lithium, dissolving alkaline earth metals and incorporating them into the electrolyte. This could be how Oxides play their role in opal formation. Oxides could come from volcanic ash, wood ash, or simply from the surrounding clays. Most of the oxides are found as impurities in opal at very minor amounts.

4.51 Ammonia & Safety

Is a major concern when dealing with Ammonia and TEOS and to ensure your own safety you should be using a Fume hood, A Hazmat suit, or similar precautions to avoid risks to your own health. Ammonia gas can be lethal. Protect your eyes, your skin, and your lungs by using suitable safety equipment.

4.52 Summary

- Ammonia acts as a catalyst in the production of Al (OH)3
- Derived from protein sources such as Dead Organic Matter [DOM]
- This decay is accelerated by bacteria & microbes
- Decay is also accelerated by Iron especially the diiron species Fe^{2+} and Fe^{3+} [Fenton Reagents]
- Ammonia also decomposes in the presence of iron.
- Ammonia can form solvated solutions, powerful electrolytes
- Aids in the forming of Silica spheres
- Ammonia and Ethanol are excreted by bacteria and microbes

Points requiring Emphasis: Len Cram associates the production of $Al(OH)_3$ with the formation of Opal, and Ammonia aids the formation of Silica Spheres. These are direct links to opal production.

Opal Precursors

4.53 Opal Precursor?

An opal precursor is basically a starting material from which opal can form. An opal precursor needs to supply liquid SiO_2. Opal is SiO_2 plus water. Nature has its ways of dissolving silica to provide the starting materials but man normally uses synthetics like TEOS for the sake of convenience.

One term which could broadly apply to all the opal precursors is the term Organosilicons, however, it seems that this term is used in industry to apply to the Silatranes, rather than to all the organic silicon species.

Here is what the experts have to say about Opal Precursors:

"Mono-dispersed spherical particles of silica (Silicon Dioxide) can be prepared from silicic acid or hydrolysis of ethyl orthosilicate and such sols can be used to make synthetic opal, since natural Opal consists of such material" (Ref: 31, pg 10).

However Leo & Barghoorn (1976) state that the main form of soluble silica in nature is monosilicic acid $Si(OH)_4$, which can be hydrolyzed from ethyl silicate (TEOS). As the concentration of silica increases, the monosilicic acid polymerizes with the formation of Siloxane bonds and the elimination of water $[Si(OH)_4 + Si(OH)_4 \rightarrow Si(OH)_3Osi(OH)_3 + H_2O]$.

As polymerization continues amorphous silica begins to precipitate with further loss of water and crystallization of the silica, opaline silica forms $[SiO_2 + 2H_2O]$.

It is interesting to note that Cellulose & Lignin are deemed to have the following formula: C_2H_5OH this is the same as ethanol. In industry they use absolute ethanol, which is about 98% ethanol [see Ch. 3 Wood Chemistry].

Scientists and Geologists tend to indicate strongly that TEOS is a synthetic product which cannot form in nature, it is only a man made product. The reasons which they cite are:

1. TEOS has never been discovered in nature.
2. The silica required to make TEOS is a very fine particle size which has been purposely been ground to this very fine size by special machinery, and it is assumed that particle sizes this fine are not found in nature.

These notions are refuted by the author. The answer to the first objection is that TEOS is very unstable in the presence of water and will decompose. Water is very abundant in the natural environment.

Part of the answer to the second objection is that man is not a creator, only God is capable of creating. Man can build or copy but not create.

The other part of the answer to silica particle size is that the silica found in rocks, clays and sands may well be too big to form TEOS but Biogenic silica has been overlooked by these scientists and geologists.

Biogenic silica is very widespread in nature being a constituent of shells, wood and bones. Many different types of ash, and even calcium carbonate also contain biogenic silica with the highest proportions being found in Volcanic ash and in wood ash with typically 70% plus. [See CH. 4]

Other reasons for refuting scientists on this matter are that I have found at least five ways that TEOS may form in nature, which all relate to the natural environment.

Leo & Barghoorn did an experiment with decaying wood in the presence of silica giving their findings as a chemical formula, which I can only conclude is TEOS. The question is why did these men leave their findings in the form of a chemical formula rather than plainly stating their findings. The answer to that question is probably peer group pressure.

TEOS does however, form in nature from both animal and plant decay.

There are other names and ways of describing TEOS which will be shown in the following pages.

4.54 Opal precursor Types

Both synthetic and natural types are:

- Silicon Alkoxides [TEOS, TMOS, Etc] Synthetic
- Polysiloxanes [Silicones] Synthetic
- Organosilicons [Carbon & Silica = Silatranes] Synthetic
- Sodium Silicate [Scott Wilson] synthetic
- Biogenic silica including Silicon Catecholates, Natural

4.55 Silicon Alkoxides

Silicon Alkoxides commonly used in industry are:

- TEOS [Tetra ethyl ortho silicate]
- TMOS [Tetra methyl ortho silicate]

TEOS is one type of Alkoxide, one type of opal precursor, which are composed of a metal oxide and an organic component usually an alcohol. TMOS is Tetra Methyl Ortho Silicate another silicon alkoxide but the alcohol component is methanol rather than ethanol. There are other Alkoxides (Titanium or Aluminium) but the only ones relevant to this study are the Silicon Alkoxides.

Do Silicon Alkoxides form in nature? Silicon Catecholates are biogenic silica but also could be classified as a Silicon Alkoxide.

Scientists and Geologists believe that this is impossible, however the author is not convinced, as it is an organic molecule, most of which are formed first in nature before mankind learned to copy them. If Opal forms in then Lab using TEOS, then it is highly likely that it also forms in nature.

Two possible scenarios for the natural formation of TEOS have been suggested by the author in chapters 3 and 4 by:

- the dehydrogenation of Ethanol
- Silicon tetra acetate formation

Both of these reactions rely upon the oxide catalysts found in ample supply in opal producing areas, and possibly playing a role in forming the opal caprock known as silcrete.

The problem with monosilicic acid is that it tends to form a very thick hard gel, with a particle size that may exceed the size needed to form opal. The thickness of the gel is also a problem as for the required reactions to take place the monosilicic acid needs to be thinned out.

The TEOS also seems to be more reactive. Or is it simply a matter of thinning Monosilicic acid and initiating condensation reaction?.

The Condensation [Polymerization] of TEOS occurs in three stages, i.e. three different reactions:

- Hydrolysis [Water Absorption][Acid or Base assisted]
- Water Condensation [Elimination of water]
- Alcohol Condensation [Alcoholysis – Elimination of Alcohol]

$$Si(OC_2H_5)_4 + 4H_2O + Catalyst \rightarrow$$
$$Si(OH)_4 + 4C_2H_5OH$$

Further polymerization then occurs

$$nSi(OH)_4 \rightarrow (SiO_2)n + 2nH_2O$$

This reaction formula shows what happens to the Silica but what about the Ethanol, does this remain unchanged or does it perhaps break-down to carbonic acid, which would dissolve more silica?

This method is often referred to as the Stober or the Stober Fink method.

The catalyst used in this method to condense the Silicic acid is Ammonia and water which causes the hydrolytic breakdown of the TEOS in a water Ethanol solution.

The water ammonia ratio determines the ultimate particle size.

See Also:

- Sol-Gel Gateway [Ref: 41]
www.lpi.usra.edu/meetings/resource2000/pdf/7014.pdf
[Ref: 42]

- Mauritz Sol Gel Chemistry [Ref: 44]
http://www.psrc.usm.edu/~mauritz/solgel.html

Generally speaking the polymerization process begins with the addition of water [water hydrolysis] to the TEOS. This reaction by itself is very slow. However catalysts are used to speed up this reaction. These catalysts are acid or base catalysts. The acids work much faster than the bases. The other reactions involved will automatically take place after water hydrolysis begins. The other reactions can occur concurrently once hydrolysis is begun and conditions are right.

> *"Mineral acids (HCl) and ammonia are most generally used, however, other catalysts are acetic acid, KOH, amines, KF, and HF." [Ref: 44, Mauritz]*

Ph required for Acid/base water hydrolysis

> *"Iler 25 divides this polymerization process into three pH domains: < pH 2, pH 2-7, and > pH 7." [Ref: 44, Mauritz]*

Another acid sometimes used is sulfuric acid. A pH above 7 is needed for Ostwald ripening.

The success of this reaction depends on using monodispersed silica spheres, in other words, many single spheres dispersed in a liquid, termed a SOL [Solution]. If the particles become crowded [aggregated] the gelling will occur and the silica spheres will not grow to the required size to form opal. Ultrasonics can be used to prevent aggregation and even break-up clusters.

For opal production the particles should only aggregate or gel after the spheres have grown to Opal size [150-400nm] by Ostwald ripening [also referred to as Aging].

If aggregation occurs prematurely then the end result will not be the highly prized precious opal but Potch [or common opal.]. The use of excessive amounts of ammonia or salt will also result in the formation of Potch as it will interfere with the proper development of perfect silica spheres and prevent the process of Ostwald ripening, thus opal spheres will not be formed or properly arranged into the closed packed silica arrays.

The ammonia source most commonly used in industry is Ammonia water 20 – 30% concentration. However this is a very dangerous chemical at this strength.

> "Ammonia is the most soluble of all gases. Ammonia dissolves in water to form aqueous ammonia solution. It is also called dilute ammonia solution, the weak alkali, ammonium hydroxide, ammonia water, and "cloudy" ammonia (a cleaning solvent). A solution in ammonia water is called an ammoniacal liquor. [NH3(aq) is used because while "NH4+" ions and "OH-" ions can be detected, "NH4OH" cannot be detected, so ammonia solution is shown as "NH3(aq) + H2O(l)"]" [Ref: 18, Wiki Ency.]

It may be possible to release ammonia gas by adding NaOH to the solution but great care must be taken as this has the potential to be very dangerous, it is an exothermic reaction, and can produce large amounts of gas with much foaming.

Ammonia function:

- Sphere controller [formation & size]
- Condensation initiator [catalyst] or,
- Base for condensation & as a
- Dehydrating agent

Ammonia can be used as a catalyst, i.e. in catalytic amounts or as a Base, for Base condensation reaction of TEOS, which will require larger amounts.

See also Chapter 4 Polymerization & Chapter 2 Carbonate Reactions.

4.56 Clues from Len Cram.

He claimed that he went home one day with some opal dirt, added some K-Feldspar some water and mixed up an electrolyte placed in a glass bottle. He then shook the bottle to ensure a good mix, then left the bottle on a shelf. In approx 4 - 6 days seams began to form within the clay.

This info suggests that he had the electrolyte at home, indicating some form the ability to source the materials from somewhere nearby.

Source: The World of Opals.

Ingredients:

- Opal Dirt = 75% Kaolin, 20% Smectite, 5% Illite
- Potassium Feldspar
- Water
- Electrolyte
- TEOS

New information from Len Cram led me to new conclusions.

In the Article "Secrets of Growing Opal" Len claims that the Opal forms from TEOS Tetraethyl silicate [Si(OC2H5)4] an alkoxide based on Silicon oxide SiO2 and Ethyl Alcohol C2H5OH. TEOS which is an organic molecule containing silica, possibly derived from plants.

4.57 Silicon Catecholates

Catechols are chemicals formed in plants being derivatives of phenols or Benzene. Tannins fall within this group. When plant catechols combine with silica they form a type of biogenic silicon catecholate, the formula for one type is (M+) $2[Si(C_6H_4O_2)3]$. Silicon catecholates could be described as silicon alkoxides, a form of organosilicon and a precursor to opal formation. The reaction of silica with catechol is as follows:

$$3C_6H_4(OH)2 + SiO_2 \leftrightarrows Si(C_6H_4O_2)3^{2-} + H_2O + H_2^+$$

High pH causes the formation of the Silicon Catecholate bond with the loss of 2 molecules of water and 2 molecules of hydrogen. Low pH has the opposite effect causing the breaking of the bonds with the release of Silica gel and regenerate catechol.

Silicon catecholates often contain a metal cation (M+) which are known to accelerate the condensation of silica above pH 7. Plant derived polysaccharides & oligosaccharides which are high in hydroxyl ions are also capable of accelerating silica polymerization from silicon catecholate.

Corriu and his colleagues studied the use of silicon catecholates as precursors for organosilicon synthesis [7014.pdf]. This represents another possible precursor to opal formation. This seems to be supported by the work of researchers at the Nottingham Trent University who investigated the effect of proteins and amino acids on biosilicification, using a Potassium Silicon Catecholate salt as the silica source. [Ref: 46].

4.58 Synthesis Ingredients:

The ingredients come from a synthetic opal method, some of the ingredients are quite natural, but the TEOS is not.

- TEOS
- Ethanol [ethyl alcohol] plant derived (organic)
- Ammonia [catalyst], or Carbamate (UREA).
    Ammonium salts.
- Water

Ammonia facilitates the formation of silica micro spheres which are seen in electron microscope photos of opal.

Carbamate (UREA = $CO(NH_2)_2$ or CO_2NH_4) can be used to

catalyze the sol/gel polymerization of TEOS.
(http://cchem.berkeley.edu/~katzgrp/papers/nature-1.pdf)

Chemical reactions often work best within a given temperature range. Some of the temperatures that I have seen suggested for artificial opal formation, at least the forming of the silica spheres are between

25 - 35°C,

This is probably because temperatures above 35°C will cause a loss of ammonia as gas.

4.59 TEOS formation

If natural Opal forms from TEOS, the question is "How does TEOS form in nature at ambient temperatures?". We already know that ethanol and silica are available but what causes these two chemicals to combine. Is there a catalyst involved. Could this catalyst be Nitric Acid, or possibly a combination of nitric acid and carboxylic acid. Think about the case of the miner's cat. It definitely is related to decomposition chemistry.

The bones also contain phosphates and calcium. Either of these could also be the catalyst. Silica definitely exchanges with these chemicals to produce opalized bones, or opalized shell. But then what of wood (calcium is also available in some soils - - this could account for the calcium LIME), it does not contain calcium. CALCIUM however is very common. Pine cones also opalize, so what is the catalyst.

4.60 TEOS production

Referring back to "Secrets of Growing Opal" in which Len Cram claims that Opal forms from TEOS Tetraethyl silicate [Si(OC2H5)4] an alkoxide based on Silicon oxide SiO2 and Ethyl Alcohol C2H5OH.

This reaction can be accelerated by the use of oxide catalysts by a process known as dehydrogenation [see CH 2 Silcrete]

It has been stated that Liquid SiO_2 can be made by mixing Silicone with water and air (O_2). This SiO_2 can then be added to the Ethanol C_2H_5OH to form the TEOS needed for making Opal. (Ref: 39). However: unlikely this possibility.

4.61 TEOS & Ethanol [Stober method for preparing TEOS]

Ethanol [C_2H_5OH] is not just a component of the TEOS formula [$Si(OC_2H_5)_4$] but is actually a solvent for TEOS. This means that it has the ability to dissolve and to thin out the TEOS.

TEOS belongs to a group of organic chemicals called ALKOXIDES, these chemicals are most likely of plant origin. It seems that TEOS is a combination of Silicon Dioxide (SiO_2) [SiO_4]and Ethanol. Ethanol is the fermentation product of sugars which have biological origins. Is it possible that Si or SiO_2 in the presence of fermenting sugars combines with the resulting ethanol through the fermentation process, resulting in the formation of TEOS [Silicon Ethoxide].

4.62 Temperature Range

Many of the chemicals involved evaporate at low temperatures, such as the Ethanol and Ammonium. TEOS has a flashpoint of 50°C and $SiCL_4$ has a boiling point of 60°C and a melting point of 70°C. This tends to strongly indicate that Opal forms at temperatures below 60°C.

Other indicators are that opal forms in arid areas where temperatures range from 0°C - 50°C. The fact that it forms underground would put this figure below 50°C. In fact in Coober Pedy people live underground because it is much cooler.

Moreover crystal formation (opal gel) has been observed forming in the temperature range 20-28°C [Ref: 34].

Constant temperatures above 100°C at pH 7 or above may result in the Silica becoming almost anhydrous, not suitable for opal formation as opal is hydrated silica, change of phase takes place.

4.63 Alternative Electrolyte

From Infracrystal research (Photonic Crystals)

Electrolyte (disperse medium):

- Water, Ethanol, Ester (fats, oils & waxes), Acetone
- Require counter ions; NH_4^+, K^+, Na^+.
[Ref: 36].

Monosilicic Acid [H4SiO4] Ortho, Silicic

4.64 General Information

In water Monosilicic Acid is soluble and stable at 25°C for extended periods of time so long as the SiO2 concentration is less than 100ppm.

The purpose of the hydrolysis of TEOS is the formation of Monosilicic acid, Silica Spheres. Opal spheres can be as small as ¼ micron in diameter.

TEOS then is only a vehicle to produce monosilicic acid from which opal forms by dehydration reaction. The condensation of TEOS produces monodispersed silica particles of uniform size, monosilicic acid dispersed in an alcohol water mix. It is then necessary to add ammonia or a mild alkali to convert these silica particles to spheres.

Following on from Leo & Barghoorn (1976), I have discovered that Monosilicic acid $Si(OH)_4$ [H_4SiO_4]can be prepared from Sodium Silicate (SS) by the addition of a little water and some Hydrochloric acid by the following reaction:

$$Na_2SiO_3 + H_2O + 2HCL \rightarrow$$

Sodium Silicate + Water + Hydrochloric Acid ➜

$$Si(OH)_4 + 2NaCl$$

Silicic acid + Sodium Chloride (salt) ➜

Silicic acid then polymerizes

$$Si(OH)_4 + Si(OH)_4 \rightarrow$$

$(OH)_3 SiOSi (OH)_3 + H_2O$ Confused yet, here's a better explanation:

The $Si(OH)_4$ Hydrolyses $\rightarrow SiO_2 + 2H_2O$

(This is an unstable reversible reaction.)

In order to prevent this reaction reversing, there is a need to use a catalyst, ammonia is probably the best option as catalyst. This is actually a step further advanced than using TEOS which has the aim of producing Monosilicic Acid.

Monosilicic acid can be formed by other methods also when Sodium Silicate (Na2SiO3.9H2O) is reacted with acetic acid (vinegar) to liberate monosilicic acid Ref 37

The conversion of Sodium Metasilicate hexahydrate (laundry chemical) to monosilicic acid occurs at low temperature (Ref 37; Iler, 180).

It is thought by some, however that Monosilicic acid is not reactive enough to be converted to opal.

Recently however, it has been discovered that the problem with this method is not the reactive nature of the monosilicic acid but that the particles produced were not uniform, and probably not small enough to act as an opal precursor. There must be some chemical reaction in nature that corrects the problem of uniformity of sphere size if opal grows from monosilicic acid in the natural environment.

The aim of using both TEOS and Monosilicic acid is to produce SiO_2 in liquid form [or soft gel] to grow opal. It has already been established by the quote at the beginning of this chapter references 9:6; this is the commonality between all opal genesis systems. TEOS condenses to SiO_2 and ethanol, whilst monosilicic acid through hydrolysis forms SiO_2 and water.

Until recently, Sodium Silicate has not been a suitable precursor to opal but the discovery of a new ion exchange resin has changed this position.

This new process I refer to as "The Scott Wilson Method", some of the details of which are recorded in CH. 5.

4.65 Natural sources

Orthosilicates or Nesosilicates (SOLUBLE - Tetravalent SiO_4)

> "Silicon prefers above all else to surround itself with four oxygen atoms as if in the orthosilicate ion (SiO4)----. This ion actually exists in water solution, forming the weakly ionized orthosilicic acid, H4SiO4, with ionization constant K1 = 2 x 10-10. The orthosilicate ion, which we shall call the silica tetrahedron,..."
> (Ref: 33).

All you can see of it are the oxygen atoms; the silicon is safe in the middle.

This group includes:

- Aluminosilicates (clays & feldspars)
- Garnet
- Topaz
- Olivine [Peridot being the Gem form].

These silicates have formulas like the salts of Orthosilicic acid, which are tetravalent silicates, as is monosilicic acid. These then are natural soluble silicates, and most likely the source of water soluble silica in nature.

The hardest part of my search has been to find out how monosilicic acid forms in nature. I have found two ways that this can happen, one is by the action of sulfuric acid on carbonates, and the other has to do with clay reactions and weathering.

It seems a reasonable assumption that if monosilicic acid can yield uniform particle sizes by a man made process, that nature itself would have a process for doing the same. This may be through Sulfonic acid or by polyamines. Sulfonic acids form in nature by the mixing of Sulfuric acid and mineral oils. Mineral oils form from the breakdown of plant matter.

4.66 Timescale

- First noticed in 4-6 days, is whisker like seams followed by colour
  In 2-3 weeks

- Finished product Approx 5-6 months,

it may however be possible to drastically reduce this time by using Polyamines as they accelerate the polymerization reaction.

A commonality that exists between those who make Artificial Opal, and the Len Cram process, is TEOS.

Therefore the same electrolytes and catalyst should work to form natural opal. It has already been demonstrated that both Ethanol and ammonia result from the decomposition of both plant and animal matter. It is then highly likely that this is what happens in nature but within the clay some further ion-exchange occurs. The Opal absorbs minute quantities of other mineral oxide contaminates, from the opal dirt. This will not happen in the Artificial process as there are no clays involved.

Sedimentation in the Artificial processes is more of a settling process much like that used in the wine industry. Natural opal sedimentation occurs within the natural sediments, which are the opal clays and also becomes the molding material on drying out.

Carboxylic acids may also be responsible for the formation of TEOS in nature via the esterification reaction or similar type reaction.

4.67 pH Indicator

In order to properly evaluate the necessary pH for opal forming reactions you will need to use Universal Indicator Papers. These are available from a Scientific Laboratory supply firm. Universal indicator papers will give you an accurate pH count from 0 –14. Don't bother with the litmus papers as they only indicate if the solution is Acid or Alkaline.

There are also pH metres which are available if you are prepared to pay a little extra and understand about buffering.

Table 12 - Sedimentation Process

Silica Spheres	Natural Sources Olivines/ Silicates Clays Biotite Mica Synthetic Sources TEOS // TMOS Sodium Silicate by Ion Exchange [Dowex] Silbond [Pre-hydrolysed TEOS; Sphere size??] NYACOL® 2040 is a larger particle (20nm) colloidal silica
Sphere Growth	Ostwald Ripening; aging; maturing Requires size range 150 – 450nm
pH	Acid/Alkaline Dependent on Chemical used
Temperature	Great Artesian Basin average daytime temp is 35C Some Imitation processes heat mix to 35C then allow to cool within range 20 – 28C
Sedimentation	Settling time where the silica spheres are allowed time to settle into their closely packed array. This is a very uniform arrangement. If this arrangement is not properly formed POTCH results. Under Silicone oil can reduce to 6 weeks.

	Len Cram method 3 months, any size. Imitation [Stober] 7 months or more; produces very small opal; to about 1cm
Dispersant	Apart from an electrolyte [solvent]; SALT may prevent Colloidal silica forming as it causes GELLING. Too much salt results in POTCH, gels before growing to opal size. Correct amount of salt is essential; I would suggest the minimal amount to maintain the SOL. SALT Gelling Agent.
Ammonia 20 – 30% Imitation	Ammonia & water regulate sphere size, so the correct ratio & concentration needed. Ratio: Water : Ammonia : TEOS 0.5-17.0 M: 0.1- 3.0M: 0.1- 0.5 M [Mole/ltr] Shortened ratio 5:1:1 to 17:3:0.5

Oxides & Hydroxides

4.68 General Information

Firstly we will consider oxides and their possible links with opal formation, later hydroxides and radicals will be discussed. Oxides must be examined as components of volcanic ash, and wood ash, and their reactions in solutions, to determine any involvement in the opal formation process.

Virgin Valley in the USA seems to place both these ash types in the opal formation environment, which must place them amongst the suspects for the opalizing fluid, or electrolyte.

Len Cram is another source that points to oxides as being the opal electrolyte, as he says that Lava contains all the necessary ingredients for opal to form. One would not expect to find ammonia in lava however it is a very rich source of oxides as can be seen from the tables at the end of this chapter.

In Virgin Valley, we know that volcanic activity caused the felling of large forest areas. This fallen timber was later covered with hot volcanic ash from volcanic eruptions. This hot ash would have caused the exposed surfaces of the timber to burn adding wood ash to the volcanic ash. As rain watered these oxides the silica began to dissolve, it probably produced an alkali silicate gel. As the timber is completely covered by deep volcanic ash the conditions become anoxic, anaerobic, causing the silica present to rapidly dissolve.

The alkali silicate gel would be diluted by further rains and the silica would begin to penetrate into the wood structure, which continues to decay, possibly accelerated by the silica. There are at least two possible ways in which TEOS may form in under these conditions.

- o There is the method suggested by Leo & Barghoorn [See CH 4]
- o The dehydrogenation of ethanol, and possibly,
- o The formation of Silicon Tetra acetate which forms when Ethyl Acetate and dissolved silica combine under the influence of oxide catalysts [soil catalysts].
- o Many plants also contain Catecholates which are a type of silicon alkoxide very similar to TEOS which could easily be a precursor to opal formation.

Another possible source pointing to oxides is the following quote:

"Hydrous Oxides dissolve clays [aluminosilicates] at 80°C"

Source of quote unknown to author. Some publications define hydrous oxides as clays, however from the above quote; the definition of hydrous oxides must be much wider, possibly including oxides from such sources as volcanic ash and wood ash.

If the above quote is correct it could produce the gels from which opal forms.

4.69 Dehydrogenation

Oxides are important in some chemical reactions such as the dehydrogenation of Ethanol where they act as catalysts. In the presence of dissolved silica, TEOS can form as Silicon Ethoxide. See Tables at end of this chapter.

4.70 Natural Oxides

Like most hydroxides, oxides tend to be caustic. Natural oxides are used as Pigments, and Fluxes for both glass melts and for ceramic glazes.

As oxides take on water, or are hydrated they become hydroxides. When hydroxides are dehydrated they revert to oxides. If oxides are completely dehydrated they become metals.

4.71 Volcanic Ash [Rhyolite]

Len Cram dropped a clue that Lava contains all that is required for opal to form. The Lava type most associated with opal is Rhyolite.

Volcanic Ash is produced by volcanoes that have magmas that are high in Silica, like Rhyolite magmas. The ash thrown out from these volcanoes is not soft like wood ash but is hard as it is a glass ash. Volcanic ash is also known as Pumicite and the larger portions, rock size bits as Pumice.

Similar in composition to Alkali feldspars, [Sodium & Potassium Feldspars] due to its Rhyolite origins.

Volcanic ash is said to dissolve easily in dilute alkaline or saline conditions. In the ground it degrades over time into Bentonite clay.

4.72 The Alkali-Silicate Reactions [ASR]

This reaction is known by road makers worldwide, as it often forms in cement causing cracking and early deterioration of the cement. The two main alkalis involved are Potassium Hydroxide and Sodium hydroxide which when they occur along with silica, they dissolve the silica, combining with it to form an alkali silicate gel. Other hydroxides can also be implicated such as Calcium hydroxide. As gels form they expand, causing the damage to the concrete. The only known hydroxide to have a beneficial effect is Lithium hydroxide, sometimes used to counter the effects of ASR.

This is yet another testimony of the effectiveness of the ability of oxides to dissolve silica.

The word acid used in the context of Lava is often a reference to the SiO_2 content rather than other acids. Acid rocks are rocks high in SiO_2.

Wood is often an Opalized product; wood chemistry is probably involved in the opal formation process. The similarity in volcanic ash and wood ash is quite striking. Pumice is also close to volcanic ash in composition, perhaps a combination of the two may produce the best results.

4.73 Pozzolans

There are three main types of Pozzolans, often used in concretes, these are:

- Pumicite & Pumice, a volcanic grains or rock, like volcanic ash.
- Diatomite and
- Fly ash is the ash collected from furnaces and boilers; it is the remains of solid fuel burnt to provide heat. The fuel sources are mainly wood or coal. Therefore the resulting ash is either Wood ash or Coal ash. Wood ash is the grey ash not the black material which is charcoal, which is mainly carbon.

Pumice is available as pumice stone used as an abrasive to clean old skin from the feet. Avail Carrigs Chemist Colonnades $4.85 soap bar size.

Wood ash maybe available from pottery or ceramics suppliers, or from cement producers [wholesalers]. Used by some as soil fertilizer.

As opalization is a dehydration reaction, I continued to research the effects of oxides as dehydration agents in order to discover if there were any links to the opalization process.

4.74 Oxide Catalysts

Oxides indeed are catalysts not only for dehydration but also for dehydrogenation. Oxides are commonly referred to as soil catalysts.

In an acid medium oxides tend to be dehydration agents but in base mediums oxide catalysts favour dehydrogenation.

There is quite a large list of metal oxides which act as catalysts in the dehydration or dehydrogenation of ethanol. Volcanic ash; wood ash; fly ashes; volcanic water and mineral water are all high in mineral oxides. In acidic mediums with oxides present the dehydration of ethanol leads to ethylene followed by diethyl ether even at low temperatures.

When oxides are in basic solution of ethanol acetaldehydes are formed by dehydrogenation. Some of the oxide catalysts such as Al2O3 in ethanol will form ethoxides.

Ethanol in the presence SiO_2 formed the ethoxide $Si(OH)$ $3(OC_2H_3)$, which formula is very similar to TEOS. Could this be an intermediate to TEOS formation or perhaps exists in the condensation pathway of TEOS, which forms opal under the right conditions.

This is very similar to Leo and Barghoorn's experiment (1976) in which they hydrolysed TEOS showing the formation of monosilicic acid and the formation of Siloxane bonds resulting in the elimination of water.

In fact they showed the mechanism for silica nucleation in wood and plant tissue involved hydrogen bonds forming between the hydroxyl groups and the Silicic acid and that of the cellulose and lignin of the plant tissue. $C_2H_5(OH)$ + $Si(OH)$ 4 → $C_2H_5-(OH)$ 2-$Si(OH)$ 3

Compare the two:

Ethoxide $Si(OH)$ $3(OC_2H_5)$

$C_2H_5\text{-}(OH)\, 2\text{-}Si(OH)_3$

It seems that the same result is reached in two quite different ways.

4.75 Silicon Ethoxide is TEOS

Ethanol is converted to the ethoxide ion $(CH_3CH_2O^-)$ by reaction with alkali metals (or their salts), but especially by metal oxides. An investigation of the Claisen Condensation reaction has revealed that ethoxides are catalysts for the decarboxylation of amino acids. The ethoxide reaction is promoted by Potassium but inhibited by Lithium.

A precursor to TEOS has been found, this discovery is that TEOS is derived from Silicon Ethoxide [Aspen Aero gels]. In fact it is not just a precursor, it is in fact just another name for TEOS [EnvironmentalChemistry.com; Ethyl Silicate Synonyms.]

Now it seems that we have a link between Plant decay, amino acid decay and metal oxides, that being Silicon Ethoxide, in the opal forming process.

The ability of Silicon Ethoxide to form in nature represents a

Possible link to opal morphology.

Common Oxide Sources

Table No. 13 -- Oxide Comparison No 1			
Volcanic Ash [Typical Analysis] Wt		**Coal Ash** Pulverized [calcined?] Wt	
CaO	6.0	CaO	0.43
K_2O	1.0	K_2O	NIL
Na_2O	4.0	Na_2O	NIL
MgO	3.0	MgO	0.52
Al_2O_3	15.0	Al_2O_3	1.19
SiO_2	67.0	SiO_2	4.73
Fe_2O_3	2.0	Fe_2O_3	0.37
TiO_2	0.5	TiO_2	0.09
		P_2O_5	0.05

Table No 14 -- Oxide Comparison No 2			
Pumicite [Larger Volcanic grains]		**Wood Ash** [varies with species]	
CaO	0.4	CaO	15.95
KnaO	11.05	K_2O	0.84
MnO	0.17	MgO	1.57
Al_2O_3	12.5	Al_2O_3	0.63
SiO_2	74.2	SiO_2	71.96
Fe_2O_3	1.6	P_2O_5	0.42

Table No 15 – Comparison Silcrete & Coal [**Kerogen**] [Oxidized by Potassium Permanganate].		
Silcrete		**Coal** + [KMNO$_4$]
Silica	85%	Carbonic Acid H$_2$CO$_3$ x 42%
TiO2 Est	10%	Acetic Acid CH$_3$COOH x 2%
Fe2O3 Est	2%	Oxalic acid HOOH-COOH x 7%
Al2O3 Est	3%	Benzenecarboxylic acid x 48%
		Succinic Acid x Trace amounts

Table 16. Major Oxides in Escott Zeolite [Clinoptilolite]

Major Oxides	Zeolite Rock (%)	Major Oxides	Zeolite Rock (%)
SiO$_2$	68.26%	CaO	2.09%
TiO$_2$	0.23%	Na$_2$O	0.64%
Al$_2$O$_3$	12.99%	K$_2$O	4.11%
Fe$_2$O$_3$	1.37%	P$_2$O$_5$	0.06%
MnO	0.06%	SO$_3$	0.00%
MgO	0.83%		

Table 17. Oxides Comparison Bentonite Clay & Potassium Feldspar			
Bentonite ActiveBond 23 [pH 10] **Feldspar** Typical Analysis		**Potassium**	
SiO_2	56.0%	SiO_2	66.6%
Al_2O_3	16.2%	Al_2O_3	18.4%
Fe_2O_3	4.5%	Fe_2O_3	0.07%
Na_2O	2.1%	Na_2O	3.7%
MgO	2.6%	MgO	<0.01%
CaO	1.0%	CaO	0.05%
K_2O	0.4%	TiO_2	<0.01%
TiO_2	0.3%		

.

Table 18. Oxides Comparison Vermiculite & Perlite			
Vermiculite Typical Analysis		**Perlite**	
SiO_2	38 -46%	SiO_2	70-75%
Al_2O_3	10-16%	Al_2O_3	12-15%
MgO	16-35%	Fe_2O_3	0.5-2%
CaO	1-5%	Na_2O	3-4%
K_2O	1-6%	MgO	0.2-0.7%
Fe_2O_3	6-13%	CaO	0.5-1.5%
H_2O	8-16%	K_2O	3-5%
TiO_2	1-3%		

4.76 Opal Trace Elements [Impurities]

The purpose of researching and comparing oxides is to try to identify the source rocks for opal formation, and they are also a major component of clays in which opal forms. The impurities in opal are oxides and have been identified by Len Cram as:

- H_2O 6.1%   Water
- Al_2O_3 1.8%   Aluminium Oxide
- CaO 0.8%   Calcium Oxide
- Na_2O 0.4%   Sodium Oxide
- Fe_2O_3 0.2%   Hematite
- TiO_2 0.01%   Titanium Dioxide
- ZrO_2 0.01%   Zirconia
- MgO 0.05%   Magnesium Oxide
- Ag_2O 0.002%   Silver Oxide
- MnO 0.005%   Manganese Oxide
- CuO 0.0008%   Copper Oxide

4.77 Zeolites

Another major natural source of oxides is the Zeolite family which are volcanic in origin, a type of hydrous aluminosilicates, there are a number of different varieties with one of the most common being Clinoptilolite.

The following table shows the oxides and the percentage by which they occur in the natural rock. Zeolites from different origins may vary slightly in composition. The example below is Escott Zeolite from Zeolite Australia.

4.78 Perlite

Perlite is another aluminosilicate similar in many respects to Zeolite.
Perlite finds a wide range of uses in industry and its uses and popularity are growing worldwide.

4.79 Vermiculite

Typical formula $(Mg,Ca,K,FeII)_3(Si,AL,FeIII)_4O_{10}(OH)_2O_4H_2O$

4.80 Earth Oxides

These oxides are also known as soil oxides and are found within soils and clays. Most of these oxides have catalytic properties. There are common oxides, many of which are ochres which are used as colouring agents. Rare earth oxides are also known. Silcrete contains the following:

- TiO_2 Titanium Dioxide
- Fe_2O_3 Red Iron Oxide and
- Al_2O_3 Aluminium Oxide

These catalytic agents may be involved in the dehydration of ethanol in the reaction that produces Silicon Ethoxide [TEOS].

4.81 Nucleating Agents

These minerals and oxides provide the centres to which soluble silica attaches in the sphere forming process. These are tiny dust like particles with an opposite charge to silica, thus attracting the silica in solution.

These minerals and oxides are:

- Zirconium
- Zirconium Phosphate
- Zirconium Hydroxide
- Aluminium
- Titanium and
- Thorium

See also 4.93

4.82 Oxide Comparisons Result

However, Byron Deveson has now identified the source rocks as Carbonatites and Natrocarbonatites as the closest match by oxide comparison.

Synopsis of Virgin Valley Opal Formation

Figure 14. Virgin Valley Opal

The Volcanic activity cause seismic localised earth movement bringing down timber, becomes covered with hot volcanic ash as it lays in the valleys. The heat burning some of the timber causing wood ash to form under the volcanic ash.

These two ashes are high in silica and other oxides. According to Leo & Baughorn the timber will decay to ethanol which can form TEOS by dehydrogenation reaction with oxides acting as catalysts. The TEOS then only needs water and a form of ammonia to decompose forming opal gel. Acid Sulfites and Ammonium chloride are common in volcanic environments. The burial of the timber also accelerates silica dissolution by providing an anoxic environment, along with the ethanol as solvent and alkaline solutions caused by rain and oxides. Oxides seem to play a pivital role in opal formation.

4.83 Virgin Valley Opal

This may well be another sedimentary process, as it seems that it is not direct heat from the volcanic source that causes opal genesis in this case. Whilst the volcanoes definitely act as a source for the volcanic ash [see Table 13], which contains many oxides and very fine particulate silica. They do not directly form the opal which is a product of a sedimentary process developing over a longer period of time perhaps over 2 years or more.

Radicals

4.84 Fenton reactants

Often radicals can produce stronger reactions than most ordinary chemical reactions. There are a number of radical species which have special properties and reaction strengths.

Iron and Copper are by far the best promoters of Hydroxyl radicals by way of the Fenton reaction. Iron is most used by industry.

Hydroxyls are the unique silica solvent which comes from water HOH itself and is known as:

- Hydroxyl groups [Hydroxide] OH- & Hydrogen ion H+

Flow Diagram 2. OH- Radical [Catalytic action]

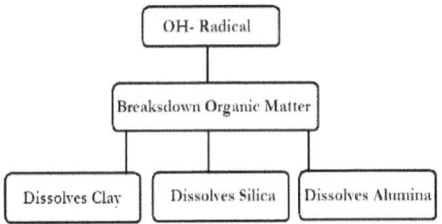

This diagram gives us an insight into the significance of the Hydroxyl radical and in the possible formation of Opal.

TABLE 19. - Role of Hydroxyl in Opal formation	
Silica	The OH- ion is the unique Silica catalyst i.e. it *dissolves* Silica [Iler].
Aluminium	The OH- ion Hydrates the Al_3+ ion to form $AL(OH)_3$ [Aluminium Hydroxides] Gel.
Ammonium	The OH ion reacts with NH_4+ ion to produce NH_3 Ammonia gas or liquid in water, at pH 9.
Carbonic Acid	Splits water HOH into H+ and OH- very effectively
Ammonium salts	Hydroxyl activators; initiators opal polymerization

Table 20: SUMMARY OF REACTIVE OR RADICAL SPECIES

1. Superoxide radical	1. Fluorine 2.23	The FENTON Reaction
2. Hydroxyl radical [OH-]	2. Hydroxyl Radical (OH-) 2.06	Carbonic acid reaction
3. Hydroperoxide radical [OOH]	3. Atomic Oxygen (singlet Oxygen)	Humic & Fulvic Acids
4. Alkoxyl radical	4. Ozone 1.50	Amino Acids (Carboxylic)
5. Peroxide radical [PR]	5. Hydrogen Peroxide 1.31	Alcohols & Sugars
6. Nitric oxide radical [NOR]	6. Perhydroxyl Radical (OOH) 1.25	Advanced Oxidation Reactions
7. Singlet oxygen [?]	7. Chlorine Dioxide CLO_2	Sulfate Radical Reactions
8. Hydrogen peroxide [H_2O_2]	8. Bromine 0.80	
9. Hypochlorous acid [HOCL]	9. Iodine 0.54	

Table 20 Continued

Methods of Generating Hydroxyls	Chemicals Enhance Fenton Reaction	Associated Chemical Reactions
Fenton & Sulfate Radical Reactions	* Alpha-hydroxy acids AHA's	Photo oxidation with oxide catalysts
UV irradiation of water	Lactic acid	
Radiolysis	Glycolic acid	
X-ray	2-hydroxy isobutyric acid	
Gamma ray irradiation	*FeII-Oxalates	
Ultrasound	*FeII-Citrates	
Hot Water Treatment [HWT]	*Tartaric acid	ROS breaks down organic matter
Conditions suitable for Fenton reaction: Temperature range 10-20C with pH at 2-4 acidic		
* AHA related		

4.85 Analysis of Radicals

The OH- ion is critical in the formation of opal, as it is one of the few ions capable of dissolving silica effectively and quickly. Iler refers to it as 'the unique Silica catalyst'.

4.86 Sugars Hydroxyls

Sucrose [Table Sugar] contains 8 Hydroxyl groups. Other sugars or carbohydrates may contain similar numbers of hydroxyl groups. Therefore sugars may well have an impact on opal formation.

Polymerization Basics

4.87 Initiators

If you have researched polymerization initiators you would
know that the most common initiator used in industry is
Hydrogen Peroxide (H_2O_2) in its various forms. It is very
common in photochemical reactions.

Another common commercial catalyst used is Phosphoric
Acid.

Conventional science has been unable to account for the
natural formation of opal but this has not stopped them from
putting forward their ideas as though they were proven facts.

The problem is that the ideas put forward by these scientists do
not line up with the facts of opal formation. Conventional
science has not put forward Ion exchange or Polymerization as
being the main processes involved. Nor has it put forward a
credible timeframe.

4.88 Hydroxyl Sources

Bases such as the Alkali's are hydroxyl donors, even more so
the hydroxides. Ethanol may also donate hydroxyl groups
which could combine with some of the Alumina to form
Aluminium Hydroxide, or with the Potassium in the Feldspar
to form Potassium Hydroxide.

Acetic Acid also contains some hydroxyl groups. The hydroxyl
group is considered to be a unique silica catalyst. Sugars and
carbohydrates are also high in hydroxyl groups. Carbonic acid
however must be considered the master of Hydroxyl
formation.

4.89 Water Polymerization

The main focus of the following polymerization information (terminology) is centered on the water polymerization processes rather than silica polymerization, but is helpful in understanding the reactions.

Gelation through polymerization is followed by dehydration of the electrolyte fluid bringing about gelation through dehydration the second stage of opal gel polymerization.

> *Gelation through polymerization. The process of linking (bridging) sol particles through the polymerization of surface atomic bonds. O2- : oxo-bridges, OH- : hydroxo-bridges, H2O : aqua-bridges. (ref: 58; Forming Ceramic Green Body).*

Hydroxides form OH (hydroxo-bridges) groups

> *The second stage of polymerization is called Gelation through dehydration (H2O - based gels). This is a process in which the energy barrier for gelation is lowered through the absence of water molecules, frequently accompanied by an increase in electrolyte concentration at sol particle surface and a decrease in the double layer thickness. (ref: 58; Forming Ceramic Green Body).*

As yet it is uncertain if this type of polymerization is involved in natural opal formation, as TEOS is not supposed to form in nature. It is possible then that a different type of polymerization mechanism occurs in the polymerization of monosilicic acid, but is uncertain. However even monosilicic acid may need the hydrolysis step before the dehydration step.

Salts disrupt this double layer and break the colloid bonds.

Len Cram states that if you can combine two Aluminium atoms with three oxygen atoms, and if all the other conditions are right, you will form opal. Let's have a look at some ways of doing this.

4.90 Oxidation of Aluminium

Let me explain this a little more clearly. The Aluminium in the clay exists in the singlet atomic form, not being combined with oxygen. The combining of these atoms in the configuration already mentioned forms Aluminium Oxide (Al_2O_3).

This process involves silica being released from the clay by Acetic acid and possibly to some extent by the sulfuric acid. At the same time the Alumina is being released from its bonds with the silica. As the Ethanol begins to sour, from oxidation, oxygen is released which now combines with the Alumina forming Al_2O_3. The oxidation reaction can be helped along by the addition of catalytic amounts of an oxidizing agent such as Potassium Permanganate. The reaction that forms opal is also involved in the formation of Al_2O_3. Iron oxide reacts with Al to form Aluminium Oxide and metallic Iron.

$$Fe_2O_3 + Al \rightarrow Al_2O_3 + Fe$$

This reaction however is normally an exothermic one with the release of extreme heat, approx 2300°C. This explanation however is improbable; a more practical explanation is provided under the heading ANALYSIS.

4.91 Copper Catalyst

Copper metal can be used as a catalyst in water for the formation of Aluminium oxide. You may know of the experiment using Al foil which is placed in water and placing a copper coin on it. After one or two days you will find that there is a hole in the foil where the copper coin was sitting and that the Aluminium has gone into solution with the water. This is the process of corrosion or oxidation. The oxidation of metals forms the metal oxides.

This same process may also work with clays, oxidizing the Al which releases Al_2O_3 into solution, breaking the bond with the Si, and allowing the silica to dissolve much more easily.

Thus the Copper metal could the elusive electrolyte that we have been looking for to catalyze the opal reaction. Temp 290°C Approx.

4.92 Aluminium Hydroxide

It has already been mentioned that Aluminium has an effect on the polymerization of silica, and in fact Aluminium Hydroxide is used as a gelating agent. *It is also present in all three of the clays that constitute opal dirt.* The following study explains the effect of aluminium on Silica polymerization.

"Effect of aluminium on the polymerization of silicic acid in aqueous solution and the deposition of silica

The effect of aluminium on the polymerization of silicic acid was studied at pH 7, 8 and 9 in the aluminium concentration range of 0-26 ppm (Al) by spectrophotometry, gel chromatography and /sup 27/Al NMR.^Retarding and accelerating effects of aluminium on the growth of polysilicic acid particles and on the reaction between monosilicic acid and polysilicic acid were observed by changing the pH.^It is suggested that the accelerating effect on the reaction between polysilicic acid particles is due to the formation of aluminium hydroxide on the surface of polysilicic acid.^The rate of decrease in the monosilicic acid concentration in the presence of aluminium was faster than that in the absence of aluminium at pH 9, because monosilicic acid could be adsorbed rapidly on the aluminium hydroxide.^From the results it is presumed that the formation of aluminium hydroxide on the solid surface may accelerate the deposition of silicic acid from geothermal water" (ref: 56; Yokoyama, T.; Takahashi, Y.; Tarutani, T.).

Aluminium Hydroxide is a known inorganic gelator (polymerization initiator) that is capable of polymerizing Silicic acid at pH 7.5 - 10.5.

The definitions below give proof that Aluminium Hydroxide maybe present in all of the clays that constitute opal dirt; due to this fact it cannot be ruled out as an initiator in the polymerization process from which opal is formed.

Hydroxides degrade to Oxides by dehydration. This may help to explain Aluminium Oxide formation. Hydroxides form by the hydration of metallic ions in water, firstly forming oxides, then as they become saturated with water they become hydroxides.

Colloidal Process

4.93 The Colloidal Process

The process known as the Sol-Gel route or the Colloidal process provides further evidence of how silica polymerizes. Silica polymerization will more often than not follow the Sol-Gel route. Firstly, a saturated solution of soluble silica and water must be formed, normally alkaline. The sol or colloid must be kept under pH 7 to avoid gelling. Naturally, if gelling is required the pH of the sol is raised above pH 7 by using an alkaline solution [Feldspar should do it].

Polymerization will most likely start with a hydrolysis reaction. Hydrolysis is the reaction that dissolves and releases this dissolved silica for further reaction, sometimes referred to as silicification. Hydrolysis can happen without catalysts but is much faster with them. The most common of these catalysts are:

- mineral acids, HCL or sulfuric acid
- Ammonia
- Acetic acid
- Potassium Hydroxide
- Amines See DEAMINATION Aminos
- Potassium Fluoride
- Hydrogen Fluoride

Hydrolysis leads to condensation [DEHYDRATION] polymerization. This may be an alcohol producing or water producing condensation reaction.

According to Iler, Sol-gel polymerization occurs in three stages:

1. Polymerization of monomers to form particles
2. Growth of Particles
3. Linking of particles into chains, then networks that extend throughout the liquid medium, thickening into a gel.

Another description is given by Byron Deveson. The stages are as follows:

1. Sphere Formation
2. Sphere Growth
3. Precipitation of Spheres: may be caused by any one of the these:
   - Lowering of pH.
   - Lowering of Ionic Strength.
   - Lowering of Temperature.
4. End to End bonding due to van der waal's forces forming chains.
5. Chains then form bundles
6. Cross linking of bundles forms the cubic packing arrays that is the structural pattern of precious opal.

Polymerization of Silicic acid:

$$Si(OH)_4 + Si(OH)_4 \rightarrow$$

$(OH)_3\ SiOSi\ (OH)_3 + H_2O$ a better explanation below:

The $Si(OH)_4$ Hydrolyses $\rightarrow SiO_2 + 2H_2O$

This is an unstable reversible reaction. Something must happen to prevent this reaction reverting back to $Si(OH)_4$.

Opal often occurs in seams. It has been proven by Len Cram

of Lightning Ridge that the Opal actually grows in seams. This is proof of the polymerization process.

It is evident then that the assemblage of Opal is a polymerization process. The main process for the formation of opal happens within the sediment. In fact it happens in two stages. In the initial polymerization stage only a small amount of opal gel is produced.

TEOS Polymerization

4.94 TEOS Reactions

This is a known type of silica polymerization process used by industry and used as representative of silica polymerization. TEOS then undergoes hydrolysis from the water which is a byproduct of the esterification process. Hydrolysis leads to Alcohol condensation polymerization, producing monosilicic acid and Ethanol. This reaction continues as a water condensation polymerization reaction which forms silica spheres of uniform size and water, opaline silica.

TEOS HYDROLYSIS REACTION

(Gelation of Alkoxides by Hydrolysis/Condensation reaction)

$$Si(OC_2H_5)_4 + H_2O \rightarrow SiOH_4 + 4C_2H_5OH;$$

$$SiOH_4 \rightarrow SiO_2 + nH_2O$$

[See: Ref: 41, Sol-Gel Gateway]

This reaction normally catalyzed by ammonia and possibly Aluminium Hydroxide.

This is a strange reaction in that it involves two opposite reactions which work together to achieve polymerization of the silica.

The hydrolysis reaction is the lead into the polymerization reaction. Water replaces alkoxide groups (OR) with hydroxyl groups (OH). Even before the hydrolysis reaction is complete, the dehydration reaction [Condensation] begins resulting in the elimination of water or alcohol from the silica.

The problem with the above scenario is that opal is known to form in association with water, yet the esterification reaction is inhibited very strongly by even minute amounts of water.

Natural Opal formation is believed to form in a watery environment therefore basically ruling out the formation of TEOS in any case.

It must then be assumed that opal forms from Monosilicic acid rather than TEOS. The dissolved silica comes from the action of Carbonic acid which not only has the power to dissolve the silica, but acts to hydrate it, that is initiates the hydrolysis reaction by creating hydroxyl groups in the water [Carbonic acid forms by stealing a hydrogen ion from the water, thus leaving the OH group].

Len Cram gave us a clue about the polymerization process, he claimed Opal would begin to form in the glass jars but only about half would turn to opal. For the remainder of the opal dirt to form opal it was necessary to drain off the liquid, the electrolyte, from the glass jar and the remaining opal dirt would become opal gel with a cap on top. My deductive reasoning suggests that this cap would be a type of silcrete, as this is what happens in nature. Most sedimentary opal is found with a silcrete capping, however the silcrete capping is sometimes displaced but will be found in the near vicinity. It is for this reason that silcrete is considered to be a marker for those looking for opal.

4.95 TEOS Polymerization formula:

$Si(OC_2H_5)_4 + H_2O$ + Catalyst [Ammonia or its salt]

$Si(OH)_4 + Si(OH)_4 \rightarrow Si(OH)_3 Osi(OH)_3 + H_2O + C_2H_5OH$

SiO_2 n H_2O [Opal gel] + $H_2O + C_2H_5OH$

SiO_2 n H_2O water & ethanol are evaporated, allowing the opal gel to dry to an opal solid.

Compare also with Silicic acid & Sodium silicate in Chapter 4.

4.96 Silica Spheres [Byron Deveson]

The method by which silica spheres arrange themselves is not yet fully understood, however current thinking tends to center around ionic charges and van der Waal's force of attraction.

Solutions with low salt content, and pH above 7, it is thought that silica spheres are repelled by each other due to the electrical charge on them. The strength of this charge being determined by the ionic strength of the surrounding fluid, and the pH of the solution.

When the pH of the suspended silica spheres is lowered, the spheres can more easily approach one another due to a drop in the overall electrical charges on each sphere. As the concentration of the silica spheres increases so too does the attractive action of the van der waal's force. The electrical charges on each sphere tend to favour end to end bonding creating linear chains. As the solution becomes more concentrated these linear chains form linear bundles or rods because this formation reduces the repulsive electrical forces between linear chains.

It is expected that further concentration would cause van der waal's forces to dominate resulting in crosslinking of the chains. This is a form of polymerization. The crosslinking would cause a cubic packing arrangement which is known to exist in precious opal. However any rapid changes in physical or chemical properties of solution at this stage would result in random crosslinking and a disordered array of silica spheres that make up potch opal.

4.97 Photonic Crystals

The old view that opal is not crystalline is now known to be false. Most of us are familiar with the typical straight edged crystals. However recently an entirely different type of crystal has been discovered. This newly discovered crystal structure is composed of spherical compact particles tightly packed in an ordered array, photonic crystals. Due to their unique spherical shape these crystals allow light to scatter in all directions. The white light that enters these structures is broken down into its inherent colours, just like a prism, only all colours can be displayed.

4.98 Nucleation

If you have been researching photonic crystal formation you may have come across the term nucleation. In simple terms nucleation is the birth of a new crystal. Nucleation is paramount in the formation of opal, for without crystal, there is no opal. (See: Ch 5 Len Cram).

As already mentioned above opal consists of micro crystals. These crystals have been observed forming in the laboratory and require temperature beginning at 35°C then dropping to between 20-28°C.

The 35°C temperature most probably related to the ammonia which catalyses this reaction. At 35°C ammonia becomes a gas.

Nucleating agents may be inorganic chemicals as listed in Chapter 2, or they could be nannobacteria. Polymorphism tends to suggest that they are contained in wood, bone, and shell. Possibly Biogenic silica which is contained in all these materials.

4.99 Potch V's Opal

You may have asked yourself why there is such a difference between precious opal and common opal or potch. The difference lies in the structural makeup of the opal. Precious opal as already seen is made up of perfectly shaped spheres arrayed in a three dimensional pattern. The structure of potch is different, the spheres may be deformed or the pattern is disrupted [see Micrograph images].

Figure 15. [Deleted]

This Figure has been deleted by the author for copyright issues.

Permission to use these photos has been requested but there has been no response to this request at this stage. It is important to the author to get this book published, so I cannot wait for these permissions to be granted.

It is hoped that permission to use these photos may be received before the printing of the 2nd edition. The micrograph photos show the difference between common opal and precious opal at a magnification of 40,000x. See Link:

Http://www.cooberpedy.sa.gov/page.aspx?u=217

"Opal occurs in many varieties, two of which are precious opal and potch. Colour in precious opal is caused by the regular array of silica spheres and voids diffracting white light, and breaking it up into the colours of the spectrum. The diameter and spacing of the spheres controls the colour range of an opal.
Small spheres (approx. 150-200 nm; I nm = 1 0'9 m) produce opal of blue colour only, whereas larger spheres (350 nm) produce red colour. Opal with red colour can display the entire spectrum.
Opal colours also depend on the angle of light incidence and can change or disappear when the gem is rotated."[Ref: 12; Coober Pedy Opal]

4.100 Developing Silica Spheres

Note in the above micrograph images that the silica spheres seem to have an irregular shape, and the packing structure of the spheres is also irregular. When you have irregularly shaped spheres you would expect that they would not pack well.

The spheres in the precious opal structure are however regularly shaped and closely packed forming a 3D array ordered of spheres.

Quickly recapping we find that the foundation of opal is obviously the Silica Sphere. There are a number of things that are important about them. These are:

- Size
- Shape
- Arrangement

The size is very important for the silica spheres to operate as photonic crystals. If they are too large or too small they will not act as photonic crystals.

If the shapes of the spheres are deformed in any way, they will not form the proper photonic blocks and will simply become potch opal of very little value.

Colloidal polymerization of silica in the opal formation process requires low levels of salts [Acid sulfates & chlorides] at a solution volume of up to 3%. (Iler, 399).

Most colloids differ in this respect to silica. In general colloidal particle growth by the Ostwald ripening technique (aging) require the total absence of salts, as they would prevent particle growth altogether resulting in a solid of much smaller particle sizes. Growth is also affected by the presence of amino acids containing nitrogen, causing larger particles to form in contrast to both hydroxyl and hydrophobic groups which limit silica sphere growth [Ref: 37].

Silica is different in that it has a double bond which requires salts to break before Ostwald ripening can occur. Too much salt however would have the opposite affect inhibiting particle growth.

The arrangement of the particles has to be a three dimensional array, so that they operate together as photonic blocks, capable of bending light in all directions. It is this ability to bend light in all directions that enables opal to separate light into all its many different colours, not just the primaries but all the hues in between as well.

How then do you control the shape and size of the silica spheres that will develop into the opal within the clay source materials?

The answer is by using the same chemical that nature uses to do the job.

These chemicals are:

- Ammonia, is used in biogenic production of opal, it is also used in man made opal.
- Shape control can be achieved by using synthetic organogelators, polypeptides, polyamines (ammonia derivatives).

Some sources suggest that the particular photonic crystals that constitute opal are in fact Christobalite, which is one of the polymorphs of silica.

When talking about sphere size in opal, we are talking about micro crystals which are measured in Nanometers, Angstroms. To view these crystals properly you need an electron microscope, because of the microscopic size of the spherical crystals. Micrograph images are photos taken using an electron microscope.

4.101 Amino acids

Despite the fact that the following chemicals have been used with TEOS there is no evidence to indicate that they would not also work with dissolved silica.

Recent research in the area of Silicification and Biosilicification has revealed that there are a number of amino acids which are thought to be active in silica condensation causing silica spheres to form; these are:

- Lysine PLL (poly-l-lysine)
- Histidine
- Arginine
- Cysteine
- Proline and
- Serine

These amino acids are thought to be more active in the form of Homopolymers. It is now known that poly-l-histidine (PLHis) enables the formation of silica spheres 150-200 nm size range at close to neutral pH. [See Chapter: Proteins, Amino Acids & Enzymes].

Other chemicals that facilitate the formation of silica spheres at pH 6.0 and at ordinary temps are [water absorbents?]:

- Poly(allylamine hydrochloride) (PAH)
- Polyacrylic acid (PAA) immobilized on substrates.

4.102 Voids

Due to the fact that silica spheres are round it is impossible for them to pack together without leaving spaces between the spheres. Some sources indicate that this space is filled with air; others think it contains water yet others believe that it is filled with silica of much smaller sphere diameter.

However recently it has been discovered that these spaces are filled by Zirconium hydroxide.

4.103 Why does Potch form?

It is possible that EXCESS salt NaCl may inhibit the formation of regular silica spheres, resulting in spheres that are irregular i.e. Mis-shapen.

If we know why potch forms we should be able to avoid its formation and produce precious opal much more efficiently.

Potch forms rather than precious opal for a number of reasons, these are:

- Incorrect ratio Ammonia: Water
- Sedimentation time was too short and the spheres have not had enough time to arrange themselves in the proper manner.
- The temperature may not have been sufficient for the crystal spheres to form properly [35°C -→ 28°C needed]
- Too high concentration of salt NaCl, best below 3% concentration.

4.104 Size & Growth

Scientists use three units of measure to measure very small things. These units of measure are:

- Nanometer [nm], 1 nm = 10 A, 1cm = 10,000,000nm
- Angstrom [A], 1cm = 100,000,000A
- Micron [u or mu], 1cm = 10,000mu

Opal sphere sizes are normally measured in nanometers

however occasionally you may see them measured in Angstroms. Opal spheres are normally range between 150-700nm.

Silica particle sizes do vary even in dissolved silica which you would expect to have uniform particle sizes but does not. It is only the very fine particle sizes that are suitable for opal formation but they must be uniform. It has baffled me for a long time as to how dissolved silica can have different particle sizes; it just didn't make any sense to me. However when I discovered that Silica has a double bond it began to become clear that the particles although dissolved that some of them were still bonded together in possibly two or three sphere groups [clusters]. This is where chloride salts can be used to break these bonds resulting in single spheres, dispersed in the solution.

In industry silica can be crushed down to very fine sizes referred to as finely divided silica powders. Nature however uses different methods to ensure that the silica source has uniform small particle sizes. One method is by volcanic eruption which causes silica to be ripped apart by Carbonic acid gas under pressure forcing this volcanic ash out through the vent of the volcano.

Biogenic silica results from the action of biological systems on silica reducing particle sizes in order to ingest the silica to incorporate it into its structures mostly as a strengthening or reinforcing agent in wood fiber, cellulose, in leaves, grasses, seeds and grains. It is used to strengthen shells and exoskeletons of crustaceans, bones in animals. The exact methods by which biological systems break down silica particles are as yet unknown. We do know that Sulfonic acids aid in the breakdown of silica particles as it is the active ingredient in Sulfonic acid ion exchange resins that are used to produce very fine uniform silica particles to produce a pure Silicic acid by removing the Sodium oxide [NaO].

The hydrolytic breakdown of TEOS may also breakdown the silica particles producing a uniform silica particle size suitable for opal synthesis. The chemical reactions that produce TEOS may also happen in the opal environment, such as the formation of Silicon Ethoxide.

There are a number of factors which will affect particle size but the most significant one is Oswald Ripening (Iler, 74). When the Opal Gel forms if the temperature and pH are right, then the silica spheres will grow. The size of the particles to some extent will influence the colour of the precious opal.

> *"Growth occurs primarily through the addition of monomer to the more highly condensed particles rather then by particle aggregation. Due to the greater solubility of silica and the greater size dependence of solubility above pH 7, particles grow in size and decrease in number as highly soluble small particles dissolve and reprecipitate on larger, less soluble particles. Growth stops when the difference in solubility between the smallest and largest particles becomes indistinguishable. This process is referred to as Ostwald ripening. Particle size, is therefore, mainly temperature dependent, in that higher temperatures produce larger particles….[Ref: 55; Mauritz, Sol Gel Chemistry; pg 6]".*

Polymerization of opal could then be described as a nucleation process as the opal seams form by this particle growth mechanism.

The nanometer size range of the silica spheres formed are within the size range of the wavelengths of natural light. White light or Visible light extends over the range of wavelengths from 400 - 700 nm (this approximates the range from the size of a molecule to that of a protozoan). Most solar radiation falls within the visible range, although our eyes view this light as all the colors of the rainbow.

Ultraviolet falls within the range 10 - 400 nm (about the size of a virus).

Beyond the 700nm lies the Infrared range up to about 1mm (Ref 5:10).

Figure 16. Silica Sphere Growth; by Author See also Iler, The Chemistry of Silica… [174] 1979.
Refer also ChemBioChem 2003, 3 pg 2 Biogenic Silica Patterning, Lopez & Coradin, Figures 1 & 2.

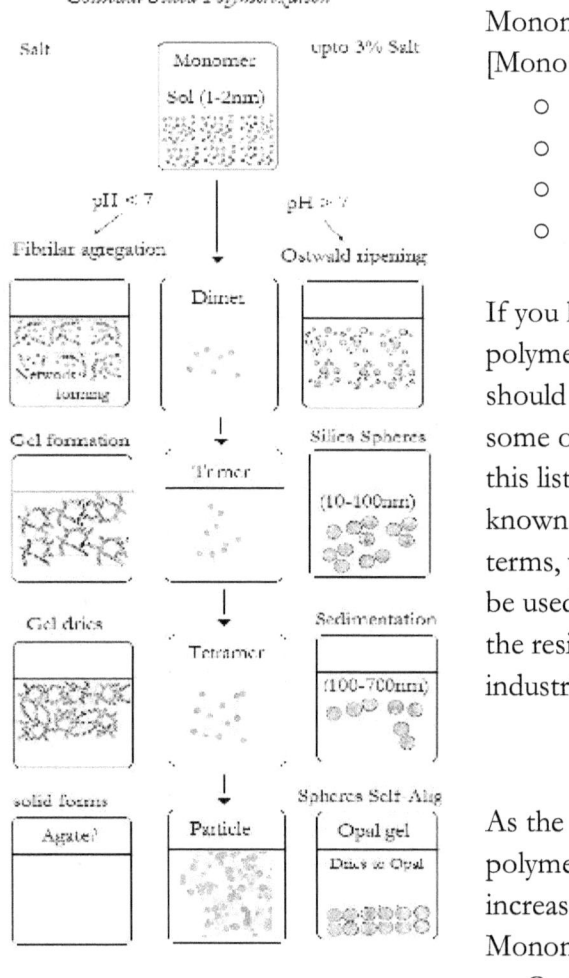

Monomer
[Monosilicic acid]

o Dimer

o Trimer

o Tetramer

o Particle
[1nm+]

If you have studied polymerization you should recognize some of the names in this list. They are well known polymerization terms, which would be used extensively in the resin and plastics industries.

As the Si(OH)$_4$ polymerizes it increases in size from Monomer → Dimer → Cyclic → Particle.

These particles in the absence of salts grow in pH 7 to 10 to become Silica Sols.

The monomer in the list is of course monosilicic acid, or Silicic acid for short. As you can see from the diagram that this is a growing network.

"For precious opal the sphere size ranges from approximately 150 to 400 nanometres producing a play of colour by diffraction in the visible light range of 400 to 700 nanometres."

(Ref: 70; How is Opal Formed? The Geology of Opal)

4.105 Viscosity problems

Most polymers including silicates have a viscosity problem, the particles tend to stick together, and although they are in liquid form a certain amount of networking has already begun.

This is why silica gels are often hard rather than soft. Preventing particles sticking together will produce lower viscosity, better flow, and produce soft rather than hard gels.

MAGNESIUM could be the solution to this problem as it prevents water from becoming a network former. This has the effect of keeping particles separate. It is yet to see if this will be as effective as using Sulfonic Ion Exchange resins. This hinges on weather the particles are actually different sizes or just beginning to network, and are sticking together, just as the author suspects.

What variety of magnesium to use? The variety of Magnesium to use is of little importance as even the carbonated variety retain this ability to thin liquid polymers and silicates [Ref: 72]. The Magnesium and carbonate ions inhibit opal formation by changing silica phase from Opal A to Opal CT. These ions can be removed from solution by montmorillonite clay through ion exchange.

4.106 Colour Control

Foreground colour is controlled by the size of the Silica spheres. Sphere size itself is controlled by the ratio of Ammonia to Water.

The background colour such as white or black still remains a mystery; possibly due to the presence of pigments such as Titanium Dioxide (white) and manganese dioxide (black). Both of these pigments act also as catalysts for Fenton reactions.

The other more credible options for background colour seems to favour Calcium Carbonate for white and Carbon for black. Opal dirt containing Calcium Carbonate is quite common and at Lightning Ridge it is called White Horse. The Carbon would come from decayed organic matter as with the Ammonia.

Opal Fluorescence

4.107 Fluorescence

Opal fluorescence has recently been attributed to the unique properties of photonic crystals [Opal spheres]. There are a number of possibilities that may explain the fluorescence in opal and these are:

- Sulfuric Acid.
- Phosphorylation (Amino Acids).
- Fluorescence exciters
  Solar UV light   (Info in Silica folder).

  Radiation: Uranium or Thorium, these are common trace elements.

- Crystal size, Fluorescence bandwidth

New studies have proven that Uranium is a major cause of fluorescence in opal even finding that opal contains tiny fragments of Uranium. [Ref: 76]

4.108 Crystal size

Small crystals of semiconductors, those in the 2- to 20-nm range, are termed nanocrystals, or quantum dots. They have different optical properties from the bulk material. As more atoms are added, the properties of the crystal vary, approaching bulk values. In particular, the band gap increases with size.

Nanocrystals of semiconductor materials such as silicon, indium arsenide and cadmium sulfide will fluoresce, and, because their band gaps vary with size, it is possible to size-tune the fluorescence wavelengths.

[ref: 71]

Another factor besides the Ammonia: water ratio that affects the foreground colours is the temperature of formation. The higher heats favour the larger size crystals resulting in red, the most sought after colour being one of the rarest. Normal temperatures in the outback can extend up to 50°C, but temperature of some rocks under solar heat have recorded temperatures as high as 80°C.

Analysis

Al Hydroxide formation from Carbonic Acid hydrolysis of Clays

The most probable way that Aluminium hydroxide is formed from the clay structure is by the action of Carbonic acid dissolving the silica releasing the Aluminium . The Aluminium then combines with water, hydrating the Aluminium to form Aluminium hydroxide:

The dissolution of silica by Carbonic acid releases Al^{3+} which is precipitated as Aluminium Hydroxide, [Refer Ch. 2 figure 6, and heading Ch. 2 Carbonic Acid and Dissolved Silica].

The Al_2O_3 Aluminium oxide for the dehydration of the opal gel is formed by the dehydration of the Aluminium hydroxide.

So apart from the clay the only other ingredients required are:

- Ammonia
- Carbonic acid & Water

TEOS formation in nature see CH's 2 [silcrete] & 4 [Oxides]

4.109 Environmental Temperature

Environmental temperature is not a major concern as the average temperature at the earths surface is 15°C. Temperature increases at approximately 30°C per kilometer. At a depth of 3 Kilometres should be approx 105°C. As you can see these are very mild temperatures. Most opal would be found in near surface conditions, below 1Km would be rare (ref: 57; Luigi Marini,).

The temperature in opal areas can reach 40°C -- 50°C. This is a definite clue as to the temperature required for the polymerization of silica to opal. The average daytime temperature in the Great Artesian Basin is 35°C.

As a general rule of chemistry, for every increase of 10°C the reaction rate doubles.

4.110 Gelling Agents

Phosphates are considered to be the primary catalysts used for polymerization in industry today. The Commercial catalysts are available in a number of different forms. In liquid form they are available as; phosphoric acid. In a more solid form they come prepared as phosphoric acid on diatomaceous earth, copper phosphate pellets and as phosphoric acid film on quartz.

There are of course many other polymerization catalysts; initiators or activator used in a wide range of different industries. Below is a fairly comprehensive list of both the organic and the inorganic catalysts or gelling agents used mostly for different industrial purposes.

Gelling Agents

Inorganic:

> Usually 2 phase systems

> -Aluminium Hydroxide gel; Bentonite Magma

Organic:

> Usually 1 Phase systems

> -   Carbomer; Tragacanth

Hydrogels:

> -Contain water

> -Silica; Bentonite; Pectin; Methyl Cellulose

Organogels:

> -Hydrocarbon type:

> - Petrolatum; Mineral oil; Polyethylene Gel; Plastibase
> - Animal Vegetable fats
>   -Lard, Cocoa butter

> -Hydrophilic Organogels:

> - Carbowax bases; PEG ointment

Hydrogels:

> Organic Hydrogels

> - Pectin paste; Tragacanth jelly
> Natural Synthetic Gums

> - Methylcellulose; Sodim CMC; Pluronics
> Inorganic hydrogels

- Bentonite Gel; Veegum

Compositions

Acacia   Alginic Ac   Bentonite   Carbomer

Sod CMC   Si Diox   EthylCell   Gelatin

Guar Gum   HEC   HPC   HPMC

Veegum   Maltodex   MC   PVA

PVP   PropCarb   ProGlyAl   Sod Algin

Sod Star   Starch   Tragacanth   Xanthan gum

Glycolate

Properties of gelling agents

- Colloidal Silica Dioxide pH 7.5 - 10.7

Related Topics Fermentation

Proteolysis, Ethanol, Ammonia, Yeast, Microbes, Bacteria, Decomposition, Composting, Amino Acids, Fermentation, Sugar Metabolism, Enzymes, Zymase, Brewing Salts.

Further Reading Fermentation
Website: PRINT from Website

Basic Fermentation Chemistry
http://www.vivo.colostate.edu/hbooks/pathphys/digestion/herbivores/ferment.html

The decomposition process -- Chapter 1

http://aggie-
horticulture.tamu.edu/extension/compost/chapter1.html

Related Topics Esterification

Carboxylic Acids, Ethanoic acid, Anhydrous Ethanol, Amino Acids, Proteins, Esterification, Polymerization, Fermentation, Polymorphism, Polymorphology, Decomposition Chemistry, Cellulose, Acetate, Biosilification, Biogenic silica, Bacteria, Diatoms, Amines, Phenols, Transesterification, Methanecarboxylic acid, Biodiesel, Isomeric ether,

Ethoxide, Ethanoate, Vinegar.

Biofuels
SAFF (South Australian Farmers Fuel)

13 Merchant Crescent, Pooraka

Related Topics - Ammonia

Solvated electrons; Organometallic compounds; Bioinorganic chemistry; chelates; electrolytes; chelates as electrolytes; charge transfer-to-solvent states (CTTS); Ion transport; Atom migration; Metal Ammonia Solutions

(MAS); The Birch Reduction; Electron donors (e.g. MAS); Proton donors (e.g. Alcohols);

Related topics – Opal Precursors:

Polymorphism, decomposition, putrefaction, fermentation, organic acids, polymerization, ion-exchange, tetraethyl silicate, catalyst, electrolyte, Photonic Crystals, Polyamines, crystallization, Christobalite, Infracrystallization, crystal growth, carboxylic acid group, Phosphorylation (Proteins/ Enzymes), Kinases, Orthosilicates, Silicon tetrachloride, Ethyl Esters, Propionic acid ethyl ester, alcoholates.

Esterification, Micelle.

Further Reading Opal Precursors:

- Kirk-Othmer Ency. Chemical Technology
  Vol 22

  Subjects: Silicon Compounds pg 38

  Silicon Esters pg 69.

  SL Lib Ref. 660.3 K59.5

- Silicification and Biosilification
  Part 6. Poly-L-Histidine Mediated Synthesis of Silica at Neutral pH

  By: Siddharth V. Patwardhan and Stephen J. Clarson

  Journal of Inorganic and Organometallic Polymers, Vol 13, No.1, March 2003.

Related Topics - Polymerization

Gelation; Activators; Catalysts; Initiators; Gelling agents; Phosphoric Acid; Hydroxides; Hydroxyls; Radicals; Electrolytes; Ionic polyelectrolytes; condensation polymerization; addition polymerization; cross linking; Hydrogen peroxide; Titanium dioxide; Thiosulfate; Iron sulfide; Acid sulfate; decomposition temperature; organic &

inorganic chemistry; photolysis; photo disassociation; gelling ph; Carboxylic Acids; Ethanol; Esterification; Colloid chemistry. Colour; Light refraction; Photonic crystal production; Play-of-colours;

Optics; photonic band gap; colour band gap; visible light spectrum;

Light bandwidth; Photonic & Sonic Band-Gap; structure of opal; opal photonic crystals; Fenton reaction; Photo-Fenton catalysts; Silica Spheres; Christobalite; Inverted opal photonic crystals; Silicification & Biosilicification; Diatoms;

Related Websites

Initiators

Sol Gel Gateway

Further Reading

The Chemistry of Silica: Solubility, Polymerization, Colloid and Surface Properties and Biochemistry of Silica...

David Jenkins, 03 April, 1980 Wiley Text Books.

Photonic Crystals - Molding the Flow of Light.

By John Joannopoulos

Optical Properties of Photonic Crystals

By K. SALODA

Bibliography:Chapter 4

Fermentation

1 *Nitric Acid*

EURO AMERICAN HEALTH website

How we become acid

The Development of Latent "ACIDOSIS"
(http://www.euroamericanhealth.com/how.html)

2 *Think Glycolic; Properties, Uses Storage & Handling*

DuPont Specialty Chemicals; Belle, WV 25051

Http://www.dupont.com.glycolicacid.

3. *Microbiology 315 Laboratory Manual*

Exercise #20: Preparation of Sauerkraut, Yogurt and Beer.
Http://www.towson.edu/~gekpenyo/315lab20.htm

Esterification:

4. *Extra GCSE-KS4 Organic Chemistry*

Doc Browns Chemistry Clinic

http://www.wpbschoolhouse.btinternet.co.uk/page04/OilPro
ducts/

ExtraOrganic.htm (5/11/04)

5. *McGraw-Hill Ency. Science & Technology*

Subjects: Carboxylic Acids <u>Vol 3</u> Pgs 252 - 255

Esters/Esterification Vol 6, Pgs 543- 545

6. *Modern Organic Chemistry 3rd Edition*

Anhydrous Ethanol (Isomeric Ether) Pg 135

A. Atkinson

Library Ref: A547.57

7. *Gasahol, E-85, and pure alcohol v's "Make your own"*

Http://running_on_alcohol.tripod.com/id28.html

8. *Ethanol resources on the Web: Journey to Forever*

http://www.journeytoforever.org/ethanol_link.html

Ammonia:

9. *Speleogenesis*

(Section on Carbonic Acid dissolution).

www.speleogenesis.info/archive/sg2/Klimchouk2/index_print.htm

10. *Artesian Springs in the GAB and their Characteristics*

(First page)

M.A. Habermehl

Principal Research Scientist

Bureau of Rural Sciences, Water Science Program, Canberra ACT

11. *Acid Mine Drainage: Chemistry*

Forms of Pollution (Pages 1-3)

www.cotf.edu/ete/modules/waterq/wqchemistry.html

Wheeling Jesuit University/NASA Classroom of the Future

12. *The Silicon Cycle*

http:www.lifesciences.napier.ac.uk/teaching/Env/Sicycle.html

(Pages 1-3)

13. *Lark Quarry Dinosaur Trackways*

Fact Sheet 5

Geology of Lark Quarry, Winton and the Great Artesian Basin

(Pages 1 + 2).

www.dinosaurtrackways.com.au

14. http://lancaster.unl.edu/ag/factsheets/288.htm

15. *The Missing Organic Molecules on Mars*

Subjects: Kerogen (Derived from Coal)
Amino Acids & hydroxyacids (Ruff Deg…).

Http://www.pubmedcentral.nih.gov/articlerender.fcgi?....

16. *BRACE Quality Assurance Project Plan (QAPP)May 2002*

Continuous Atmospheric Hydrogen Peroxide (H$_2$O$_2$)^1 and Formaldehyde (CH$_2$O) Measurements

http://hsc.usf.edu/publichealth/EOH/BRACE/H2O2%20& %20C2O%20QAPP.doc

17. *Survey of Fertilizers and Related Materials for Perchlorate* ClO$_4^-$

US Environmental Protection Agency [www.epa.gov]

Office of Research & Development; Cincinnati OH 45268

https://www.denix.osd.mil/denix/Public/Library/Water/Perchlorate/

Fertilizer/Fertilizer.html

18. Ammonia

Wikipedia

http://en.wikipedie.org/wiki/ammonia

19. Al3+ ion reaction Aqueous Ammonia
http://www.public.asu.edu/~jpbirk/qual/qualanal/aluminum.html

20. EDTA
McGraw-Hill Ency. Science & Technology
Vol 6, pgs 571
Vol 3, pgs 541
Vol 18 pgs 454

21. Solvated Electrons
Labor Law Talk Dictionary
http://dictionary.laborlawtalk.com/ion

22. *Elements of Organic Chemistry*
Richards; Cram; & Hammond
McGraw-Hill Kogakusha Ltd, Tokyo, Japan [1967].

23.
http://www.totse.com/en/technology/science_technology/166763.html

24. *Biogenic Silica Patterning: Simple Chemistry or Subtle Biology?*
Thibaud Coradin & Pascal Jean Lopez
ChemBioChem 2003, 3, 1-9
Wiley-VCH Verlag Gmbh & Co KgaA Weinheim

25. *Organic Forms of Soil Nitrogen* (pg 1) 3/12/2007
Http://www.up.wroc.pl/~weber/azot2.htm
Accessed 13/07/08

Opal Precursors:

26. *The World of Opals*   SL Lib Ref. 553.873 E19

Author: Allen W. Eckert
Publisher: Wiley 1997

27. *Modern Chemistry*   SL Lib Ref. S540.2

By Metcalfe; Williams; and Castka
Publisher: Holt, Reinhart, and Winston Inc., NY 1966.

28. Article: *Secrets of Growing Opal*

29. Article: *Some Aspects of Precious Opal Synthesis* [pgs 1-5]

Publisher: Scientific Centre for Applied Research, Dubna JINR
Russia
http://www.austgems.gil.com.au/opalsynth.html

30. *Potch Opal & Its Genetic Significance*

Authors: V.V. Serdobintseva; D.V. Kalinin
http://www.uiggm.nsc.ru/opal/links

31. *Introduction to Modern Colloidal Science*

Author: Robert J., Hunter.

Publisher: Oxford University Press, 1993

Lib Ref No. SL541.345

32. *Species-specific polyamines from diatoms control silica morphology*

Nils Kröger, Rainer Deutzmann, Christian Bergsdorf, and Manfred Sumper

http://www.pnas.org/cgi/content/full/97/26/14133

33. *Orthosilcates or Island Silicates*

Subject: Geology
http://socrates.berkeley.edu/~eps2/wisc/geo360/orthosil.ht
ml

34. *Mechanism of Growth of Supramolecular Crystals in Concentrated Suspensions of Mondispersed Spherical Silica Particles (MSSP)*

A.F., Danilyuk; V.V., Serdobintseva; D.V., Kalinin

International conference of Silica Science & Technology SILICA-1998, Mulhouse, France, Sept. 1998.

http://www.uiggm.nsc.ru/opal/links

35. *McGraw-Hill Ency. Science & Technology*

Subjects: Silicon Vol. 16 pgs 452-460
Carboxylic Acids Vol 3 Pgs 252 - 255

Esters/Esterification Vol 6, Pgs 543- 545

36. *Infracrystallization in monodisperse system of amorphous silica as a model of infracrystallization and growth of "photonic crystals".*

By: V.V. Serdobintseva; ICCG-13 Quoto Japan, June 2001.

http://www.uiggm.nsc.ru/opal/links

37. *Solubility, Polymerization, Colloid & Surface Properties & Biochemistry of Silica.*

Published by Wiley, New York c1979

Author: Iler, Ralph., K
[Pgs: 74 Ostwald Ripening; 174 Monosilicic Acid; 180 Sodium Silicate Hexahydrate; 399 Polymerization [low salt].

SL Lib Ref. 546.6832 I27

38. *ScientificPsychic.Com*

Antonio Zamora

Fats, Oils, Fatty Acids, Triglycerides - Chemical Structure
Carbohydrates - Chemical Structure

Amino Acids, Peptides and Proteins - Chemical Structure

39. *Making Gemstones Out of Household Kitchen Products*

Robert James FGA, GG
http://www.yourgemnologist.com/Kitchen/kitchen.html

40. *Think Glycolic, Properties,Uses Storage and Handling*

DuPont Specialty Chemicals, Belle WV

Http://www.dupont.com.glycolicacid

41. *Sol-Gel Gateway*
www.solgel.com

42. *Organic-based dissolution of silicates: A new approach to element extraction from lunar regolith. [7014.Pdf]*

www.lpi.usra.edu/meetings/resource2000/pdf/7014.pdf

Author: S.L. Gillett
Dept. Geol. Sci./172, Mackay School of Mines, University of Nevada, Reno, NV 89557

43. Size, Volume fraction, and nucleation of Stober silica nanoparticles

Journal of Colloid Science
 D.L Green,

44. *Sol Gel Chemistry*

Mauritz – Sol Gel Research

http://www.prsc.usm.edu/~mauritz/solgel.html

45. *Biogenic Silica Patterning: Simple Chemistry or Subtle Biology?*

MIMIREVIEWS

P.J. Lopez & T. Coradin

ChemBioChem 2003, 3, 1-9

Wiley – VCH

46. *Towards an Understanding of (bio)silicification: the role of amino acids and lysine oligomers in Silicification.*

Authors: BELTON David., PAINE Gary., PATWARDHAN Siddharth V., PERRY Carole C.

Journal of Material Chemistry 2004, vol 14, no 14, pgs 2231-2241 *

Oxides, Hydroxides & Radicals:

47. *A Review of Chemical Oxidation Technology*

2 The 4 Technology Solutions

http://2the4.net/html/chemoxwp.htm

48. *Gems made by Man*

By Kurt Nassau [pgs 258-259]
Chilton Book Co.

Radner PA., c1980

49. Title: *Free Radical Theory of Aging*

Subjects: Free Radicals – What are they?

Iron and Copper
http://www.smart-drugs.net/free

50. *Soil Treatment; in-situ chemical oxidation of contaminated soils* (using Hydrogen Peroxide) [Page 3]

http://www.h2o2.com/applications/hazardouswaste/soil.html

51. *A Computational Chemical Investigation of Dehydration and Dehydrogenation of Ethanol on Oxide Catalysts* [Intro; 3.1]

Paper by:

Yuji SHINOHARA; Tsuyoshi NAKAJIMA; Satoshi SUZUKI; and Hideaki ISHIKAWA.

52. *The Claisen Condensation*

Subject: Ethoxides
http://www.chem.uic.edu/web1/OCOL-II/WIN/CH23/F2.HTM

53. *In Situ Solid State NMR Studies of Ethanol Photocatalysis:*

Characterization of Surface Sites and their Reactivities

By: Son-Jong Hwang; Daniel Raftery

Publisher: Elsevier

Publication: Catalysis Today 49 (1999) 353- 361

54. *Fate of Silicate Materials in a Peat Bog*

Http://www.osti.gov/energycitations/product.biblio.jsp_id=5
539161

Polymerization:

55. *Sol-Gel Chemistry* (Pages 1-3; 6)

Mauritz - Sol Gel Research
http://www.psrc.usm.edu/~mauritz/solgel.html (1/11/2004)

56. *Effect of aluminium on the polymerization of silicic acid in aqueous solution and the deposition of silica*

Source:
http://www.osti.gov/energycitations/product.biblio.jsp?osti_i
d=5276043

Authors/creators: Yokoyama, T. ; Takahashi, Y. ; Tarutani, T.
(Kyushu Univ., Fukuoka (Japan). Faculty of Science);
Yamanaka, C. (West Japan Engineering Consultants, Inc.,
Watanabedori, Chuo-ku, Fukuoka 810 (JP))

57. Course Title: *Geochemical & Geophysical Methodologies in Geothermal Exploration*

Subject: Geochemical Techniques for the Exploration &
Exploitation of Geothermal Energy (92 pages).

Author: Luigi Marini; University of Genova, Italy.

Websource:
Http://cabierta.uchile.cl/revista/11/revisiones/1_1/a_manzell a.htm

58. *Forming Ceramic Green Body*

Wet Forming of Ceramic Powders

Sol-Gel Processing of Ceramics (Polymerization)

Gelation Mechanism involve destruction of the double layer that keeps particles apart & can include: (A & B only)

Http://www.mmat.ubc.ca/courses/mmat382/sections/cnc43 10.doc

59. *Biogenic Silica Patterning: Simple Chemistry or Subtle Biology?*

MIMIREVIEWS

P.J. Lopez & T. Coradin

ChemBioChem 2003, 3, 1-9

Wiley - VCH

60. Websites related to photonic crystals & photonic band gap

www.pbglink.com/

61. Nature Publishing group

www.nature.com/cgi-
taf/gateway.taf?g=3&file=/materials/nanozone/news/articles
/m030724-2.html

62. *Optical properties of photonic crystals*

www.directory.net/Science/Physics/Optics/Photonic_Crystals

63. *Photonic crystals explained*

www.aip.org/enews/physnews/2003/split/622-1.html

64. *Tunable Photonic Crystals. Photonic crystals affect the flow of photons in much the same way that electronic devices.*

www.aip.org/enews/physnews/2003/split/633-3.html

65. *links to photonic crystal websites*

www.jareed.com/pc/pbg.htm

66. *Coober Pedy Opal*

http://www.opalcapitaloftheworld.com.au/opal.asp

67. *Formation of Silica Structures Utilizing A Cationically Charged Synthetic Polymer* (Abstracts 2001 Academic Poster Session, Cleveland, Ohio).

Silicification & Biosilicification

Parts 2; 5 & 6

By Siddharth V. Patwardhan and Stephen J. Clarson

Http://www.eng.uc.edu/~sspatwar/Abstracts.htm   [Pages 1-4]

68. IMM Report Number 14

Recent Progress:Steps Towards Nanotechnology

By Jeffrey Soreff

http://www.imm.org/Reports/Rep014.html

[Pages 1 & 2]

69. *Star Light, Star Bright Teacher Page: Science Background*
http://amazing-space.stsci.edu/resources/explorations/light/star-light-science.html)

70. How is Opal formed? The Geology of Opal
http://www.opalsdownunder.com.au/articles/formed.htm

71. *Nanocrystals outshine Lazer dyes*; Photonics Technology World;

March 2000 Edition; Website:
http://www.photonics.com/spectra/tech/XQ/ASP/techid.787/QX/read.htm

72. *Tec-cement Reactions; can Hydraulic Cements & Geopolymers Merge?*

Website Article
John Harrison, Managing Director, TecEco Pty Ltd

73. *Vermiculite Analysis*

http://www.vermiculite.net

1 Sept 09

74. *Wikipedia*

Subject: Perlite
http://en.wikipedia.org/wiki/Perlite

1 Sept 09

75. *Lightning Ridge; Home of the Black Opal*

Subject: Particle sizes [pg 102].

Gan Bruce

Macquarie Publications P/L, 1983, Dubbo NSW Australia.

76. *Uranium responsible for Precious Opal*

http://news.softpedia.com/news/Uranium-Responsable-for-Precious-Opal-71729.shtml

Accessed 15/08/2013

5 OPAL SYNTHESIS

Forerunners of Opal Synthesis

5.0 Pierre Gilson

Just a little before John Slocum, Pierre Gilson in 1967 announced the synthesis of opal. This was a real synthetic opal as it was made of silica with water content like natural opal. The success of Pierre Gilson was probably an inspiration to others who were to follow in the synthesis of precious opal. [Iler, 404]. The Gilson synthetics were so similar to real opal that in 1980 the Kyocera Corporation (Kyoto Ceramics) of Japan were selling both black and white synthetic opals based on Gilson's process.

The Russian's were also producing synthetics for the market place at around the same time. The Russian's used their own formula and process for making synthetic opal, and have extensive scientific research into opal formation.

Gilson used hydrostatic pressure as a method of hardening his opal as he believed that this was the only way to produce even pressure all around the opal.

Pierre Gilson has succeeded in synthesising many other gemstones, including emerald, ruby, and even diamond.

5.1 Dr. John Slocum [Rochester, Michigan]

John Slocum was one of the first people to produce an extremely realistic imitation of opal; a simulant opal that was first known as Slocum Stone but was later renamed "Opal-Essence". The flash of colour was produced by metal foils within the body of the opal. It was introduced onto the market in the late 1970's. It had good colour but the large patches served to identify it as an imitation.

5.2 Peter Darragh & John V. Sanders CSIRO

These two men were instrumental in discovering the true nature of opal by viewing it under an electron microscope. They discovered that the opal was composed of minute silica spheres in a closely packed cubic array. These spheres are a type of microcrystal similar to Christobalite.

This discovery was the fuel for many to try to synthesis opal. They also tried to patent the opal formation process to protect the opal industry from a flood of synthetics, but with little success.

John Sanders was the first man to take photographs of opal spheres. These photographs are called micrographs.

Opal Formation Methods

5.3 General Information

In the latter half of the 20th Century there have been numerous attempts to imitate and synthesis opal. Needless to say, some attempts have been better than others. The two most prominent simulants these would be John Slocum and Pierre Gilson, who both have their own patented versions. There are however some quite distinct differences between their attempts at imitation and real opal, which I do not intend to go into.

The best of the synthetic opals was by the Russians who produced a realistic looking opal, which only lacked the fluorescence of the real opal.

The Russians produced a better looking synthetic opal because of their superior research, which was based upon essential natural ingredients.

The variance in the ingredients and the method by which synthetic and natural opal forms was closing. The Methods used for the synthesis of opal was evolving. The main differences that still remain between synthetic models and natural systems, seems to be in the route to soluble silica. Industry uses TEOS because it is one of the purest forms of soluble silica available to man. Experts say that there is little evidence that TEOS forms in nature. Soluble silica exists in nature as [mono] Silicic acid. This form of soluble silica should produce the same end result in the opal electrolyte, as both produce SiO_2 which with water is the main building block of opal.

Comparisons of the main methods of opal synthesis are shown in the Table in towards the end of the conclusion.

The Russian Synthesis Method

5.4 General Information

The Russian team of S.V. Filin, A.I. Puzynin, V.N. Samoilov., have made much of their opal synthesis research available to the public. These people and their method from this point on will be referred to as "the Russians" or the Russian method. The information they made available is complete with method and ingredients, and as such can be used by us as a base model for opal synthesis.

The Russians discovered a way to make opal over 30 years ago. They used a synthetic form of soluble silica called TetraEthylOrthoSilicate or TEOS. The actual method used is called the 'Stober method' or 'Stober- Fink method' [1968] which relies on the polycondensation [hydrolysis] of TEOS which results in the TEOS decomposing into SiO_2 (aq) and Ethanol.

TEOS is a type of organosilicon belonging to a class called the Alkoxides. The two most common Alkoxides are:

- TEOS TetraEthylOrthoSilicate [contains Ethanol]
- TMOS TetraMethylOrthoSilicate [contains Methanol]

There are other Alkoxides but these two predominate in industry. They are commonly used as sources of soluble silica.

Another group of organosilicons is the polysiloxanes which contain an alkane group rather than an alcohol group.

5.5 Russian Synthesis Ingredients:

- TEOS [Tetra Ethyl Ortho Silicate] $Si(OC_2H_5)_4$
- Ammonia(aq) [Catalyst 25 – 30% used in industry]
- Ethanol 98 – 100% [TEOS Solvent – Dispersing agent]
- Deionized Water/Distilled [Hydrolysis Agent]

5.6 Stober-Fink Condensation of Silica from TEOS

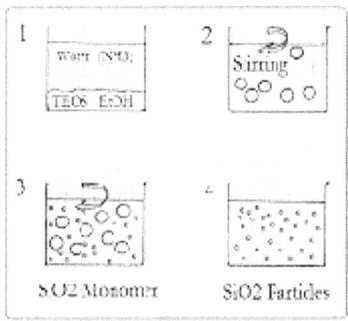

Figure 17. The Stober Method

There is a lot of variation in stober formulas as you can see by the two typical examples above. When Ammonium hydroxide is used particle size will be determined by the Ammonium hydroxide and water ratio. The critical quantity is the TEOS which should be approx. 0.5 % by volume, which allows for maximum silica content without overcrowding of the silica spheres. Excess silica leads to premature gelling without the much needed particle growth [ref: 23].

These ingredients are all placed into a reaction vessel. The first particles form at about 10nm in diameter. Growth takes place by either addition polymerization or possibly by Ostwald ripening. Addition polymerization will be dependent on how you add TEOS into the mixture, if you add it slowly or deliberately over a period of time then addition polymerization will be the growth mechanism.

If all the TEOS is added to solution at the same time then the mechanism of growth is more likely to be by Ostwald Ripening. The TEOS then undergoes a hydrolysis reaction and concurrently a dehydration reaction [Dehydration polymerization – also called condensation polymerization].

Particle growth for opals fall within the range 150-700nm approx. Most artificial or synthetic opal manufacturing processes produce sphere sizes below 500nm which accounts for the more common colours but the rarer colours, i.e. the oranges and red come from larger sphere sizes. These larger sizes can be achieved by increasing the viscosity of the electrolyte.

Viscosity is a measure of a fluids resistance to flow, i.e. water has a low viscosity whilst molasses has a high viscosity.

One chemical that can increase the viscosity of the fluid is Ethylene Glycol, however, there are probably plenty of others, and glycerin also has high viscosity as do most oils. [Ref: 23].

5.7 Sedimentation

These particles are then left to settle in the tank [or reactor vessel], a process often called Sedimentation, as the particles settle to the bottom like sediments. Whilst settling is occurring the particles are arranging themselves into a closely packed array of silica spheres. Sedimentation in the Russian system lasts for approx. 7 months [Sedimentation Ref: 1; page 2].

> *"In 1976 a team of researchers Leo & Barghoorn condensed Ethyl silicate [TEOS] with water to form Monosilicic acid [Silicic Acid], which is the main form of soluble silica found in nature. The results of their experiments revealed that as the monosilicic acid water solution became more concentrated with amorphous silica, the monosilicic acid polymerizes forming Siloxane bonds [Si-O-Si] and eliminates water:*

$$Si(OH)4 + Si(OH) 4 \rightarrow Si(OH) 3OSi(OH) 3 + H2O$$

Further Polymerization

$$SiO2.nH2O + H2O$$

As this process continues the amorphous silica begins to precipitate, and with further water loss, the silica begins to crystallize [nucleation], with opaline silica forming." [Ref: 9]

Sedimentation is sometimes called aging and it remains the most common method of collecting silica spheres and allowing them to arrange themselves into closely packed arrays for synthesizing opal.

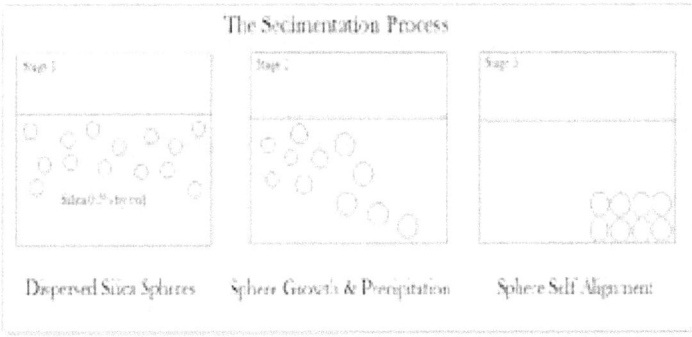

Figure 18. Sedimentation Settling Tank; showing ideal silica volume to electrolyte [Super nat]. In Stage 1 the silica spheres are single spheres evenly dispersed in the Electrolyte and liquid dispersants.
Stage 2 simply shows silica sphere growth, the larger spheres begin to aggregate and are too heavy to remain suspended in the liquid and begin to sink to the bottom. In Stage 3, Silica spheres shown at the bottom of the tank self align forming Opal Gel or Opal Cake as it is known in industry. Self

Alignment occurs due to positive and negative charges causing attraction and repulsion.

5.8 The Opal Gel Drying Problem

The Russians used a method called supercritical drying [Ref 1; page 3] which left pores in the opal stucture, which had to be filled with more silica, or other substitutes. The end result is opal no larger than 1 cm.

The Russians have improved their methods of opal synthesis since these early methods were made public.

The crazing in the opal is caused by the drying method employed. Len's method of drying although different from the early Russian method did not end in better results, as his air drying method left the opal with cracks and crazing. Cracks and crazing make even the best coloured opal almost worthless. Air drying is simply not a good option.

The process by which crazing occurs is explained very well by [Ref: 1, page 3].

5.9 Ammonia Functions & Sources

Ammonia:

- Catalyst for TEOS polymerisation [decomposition]
- Catalyses Silica Sphere growth.
- Dissolves Alkali Metal ions

Ammonia Water Sources:

- Industrial 10 – 30% [Dangerous]
- Protein decomposition

- Ammonia Sulfate/ Urea
- Decomposes in presence of iron species.

5.10 Ammonium Carbonate [Urea]

Ammonium carbonate is very unstable and decomposes to NH_3 + carbon dioxide. The enzyme, urease, greatly stimulates this reaction. <u>Urease is found in all soils in sufficient quantities to bring about rapid conversion of urea to ammonia (NH_3)</u> which then in the presence of water converts to ammonium (NH_4^+). The ammonium then attaches to the negatively charged soil particles and behaves like any other ammonium-based source of nitrogen fertilizer. This whole hydrolysis reaction normally occurs within a 2 to 4-day period. If the urea is not incorporated into the soil, hydrolysis can also occur and NH_3 can then be lost to the atmosphere through volatilization.

As seen in the table above the OH- ion which forms in alkaline conditions catalyses the conversion of the ammonium ion NH4 to ammonia gas which in contact with water becomes aqueous ammonia NH3. Urea does not form a strong electrolyte and can be eliminated as an opal electrolyte.

The two main forms of ammonia in both the volcanic regions and sedimentary settings are Ammonium Sulfate & Ammonium Chloride.

5.11 Evaluation Russian Synthesis Method

In the Russian method we can see that the opal formation process is a Sol-Gel process. Firstly a solution [Sol] is made which is later gelled, forming a jelly like substance which is hardened into the finished product.

They have provided us with an insight into:

- the chemical ingredients
- chemical process [Stober Polymerization of TEOS]

Once the Russians have produced their opals they then go through a process of Supercritical drying. This is a self defeating process which is totally unnecessary. It destroys the original structure of the opal, leaving pores within the structure which then have to be filled, and a second heating process used to bind in this new material. This process limits the size of the opal produced to approx 1cm. Hardly worth the effort.

5.12 Supercritical Drying

"As silica gels dry, they progressively shrink and harden. This leads to stress development (for non-uniform shrinkage) and possible fragmentation. The primary source of stresses during gel drying, apart from non-uniform drying shrinkage, is the capillary pressure p of water leaving pores of radius r, to result in $p = 2\Gamma/r$. For example, $p = 100$ MPa (this is large !) if $r = 2$ nm. Possible techniques to avoid cracking under high capillary pressure p include:

pore enlargement (r) (Oswald ripening $300°C$)

surface tension reduction (Γ) through use of surfactants

freeze-drying: solvent sublimation rather than evaporation *hypercritical drying- no liquid-gas interface, so $\Gamma = 0, p = 0$"*

[Ref: 17; Forming Ceramic Green Body].

Heating the Opal gel should not cause any hazards as researchers and manufacturers of artificial opal dry their silica gels at temperatures of up to 1000°C. The process is called supercritical drying. This process is probably carried out in an autoclave as references are made to pressures of 3000 atmospheres.

It is known that heating can actually strengthen the end product. At 300°C additional Siloxane bonds form within the opal structure.

In hindsight, the Russians could probably have come up with a more natural solution to the drying process.
The Russians ignored two critical elements in the natural drying of Opal.

- $Al[OH]_3$ Aluminium Hydroxide which is a natural polymerization agent for silica which dehydrates to $[AL_2O_3]$ a chemical drying agent for opal gel [opal cake as some people prefer to call it].
- Clay, acts to dry opal cake by absorbing water and by pressure applied as the clay shrinks on drying.

The Russians have done extensive research into opal formation and their knowledge and methods have advanced since these first initial attempts at synthesising opal. This information is contained in the following publications:

- Russian Geology and Geophysics
- Reaction Kinetics and Catalysis letters
- Colloidal Journal

Details for search criteria are to be found on the website:

http://www.uiggm.nsc.ru/opal/publications

TABLE 21:   Essential Russian Opal Synthesis Ingredients

TEOS [$Si(OC_2H_5)_4$]	Soluble source Silica
Ammonia [NH_3]	Catalyst TEOS Polymerization Controls sphere size/ colour
Ethanol [C_2H_5OH]	TEOS solvent; dispersing agent 99$^+$%
Water [H2O]	Hydrolysis agent TEOS condensation
These chemicals are the Benchmark [standard] for Opal Synthesis	

The Len Cram method

5.13 General Information

Len Cram is a self confessed amateur geologist who lives in one of the major Australian Opal fields, that of Lightning Ridge. Like the author of this book Len did not believe in the scientific explanation of Opal formation, or the link to the "so called" theory of evolution and its ridiculous time periods in the millions of years.

Len set out to prove the scientific explanation to be false by trying to make opal in his own backyard. In nature it is known that opal forms in opal dirt [clay]. He believed that it was the opal dirt itself which was the source of silica from which opal forms.

His experiments were extremely successful. He claims to use a special electrolyte that turns the opal dirt into precious opal. Len has not disclosed all the details, but he has left us some clues as to his method.

5.14 Len's Clues

Len has identified the ingredients needed for his method of opal formation as:

- Feldspar [varieties; Potassium; Sodium and Calcium].
- Opal clays [Montmorillonite, Kaolin]
- Alumina [available in clays]
- Water
- An Electrolyte; an ionic polyelectrolyte [conductor]
- A Nuclearite [attracts ions]

The electrolyte is the key to the whole process, the all illusive opalizing fluid. Len has kept us guessing as to the nature of the electrolyte, desiring to keep it a mystery.

5.15 Ionic Polyelectrolytes

Len has simply described his electrolyte as an ionic polyelectrolyte. This red herring has been thrown into contention in order to keep the electrolyte a mystery.

Ionic Polyelectrolytes covers both the Anionic and Cationic polyelectrolytes leading to the possibility of hundreds of thousand of chemicals. This was done in the hope that no-one would pursue this electrolyte.

5.16 Narrowing the Possibilities

There is a way to shorten this list to give only a few possibilities. Some may call this speculation but it is based on deductive reasoning and known fact.

First Possibility: Len has revealed that he uses TEOS as the soluble silica source. Deduction: This indicates that the reaction used is the well known stober reaction. Now the electrolyte is limited to the catalysts used for the stober reaction. These are both acidic and alkaline, however only the alkaline produces opal spheres. This supports the second possibility.

Second Possibility: Clues from nature:[1] We already know from volcanic opal studies that when Acid Sulfates mix with neutral chlorides that opal forms [Corbett & Leach 1998, Regolith 2004, pg 264].

[2]Similarly in Mound Springs that are highly alkaline [approx pH 8.7] and saturated with silica, when pH is lowered slightly by acidic water entering the system causing opal spheres to

precipitate [Byron Deveson, pg 5].

Knowing that Len is an amateur geologist, he would also know these things. Len has already described how he likes to source his materials from nature's vast reserves. Living in Lightning Ridge gives him access to these great natural opal resources.

Len would know about mound springs from his own research, and would know where pools of acid sulfate waters are located in his nearby vicinity. Acid sulfate pools are often identified by their green colour caused by the iron content.

Len's Electrolyte: It is highly probable that the second possibility represents Lens electrolyte. Articles about his method suggest that he mixes two liquids together in a glass jar then adds the TEOS and shakes vigorously until mixed, revealing the typical milky white colour of a TEOS condensation reaction.

These liquids then are Mound Spring water pH 8.7 approx and Acid sulfate water and TEOS mixed in the correct proportions. A pH meter could be extremely helpful in determining the correct mix.

5.17 Cram Electrolyte as a Sol

Len's process is a two pot mix. He makes an electrolyte, which is the 1st pot, and the clay [opal dirt], water and feldspar are the 2nd pot. It must be assumed that the opal dirt or the spring waters contains naturally occurring nuclearites.

It was claimed by Len that his electrolyte has an indefinite shelf life. This brings up the nature of the pH of his electrolyte, and he states that pH is significant in his process.

If the solution was maintained above pH 7, then we would expect the opal spheres to grow possibly to the extent of gelling. This would be disastrous as it would ruin the electrolyte.

In order to maintain a silica sol, the pH must be adjusted to below 7 slightly acidic which should yield a stable solution with a long shelf life.

This is achieved by mixing alkaline Mound Spring water [pH 8.7] with Acid Sulfate water until the solution is just below pH 7. A pH meter would be very helpful in making these adjustments.

After mixing the Mound spring and Acid sulfate waters, the TEOS is added and the bottle is well shaken to ensure proper mixing, the solution will turn a milky white indicating TEOS condensation, the TEOS breaks down releasing SiO_2 and ethanol.

This electrolyte solution is a silica colloid.

The next stage is to add some of the electrolyte mix to each of the opal dirt mixes. Once again, these are shaken really well allowing proper mixing to occur. In order for opal to grow, the pH must be raised to approx 8.7. It is the feldspar in the clay mix that achieves this objective, thus allowing for opal sphere growth. The heat of summer days will also aid the process of sphere growth.

Opal will begin to form within the glass containers within the clay.

5.18 The Missing Ingredients

When we check Len's list of ingredients against our base model it becomes quite obvious that a number of organic chemicals are missing from his list. So, then, how do we account for these missing ingredients.

Microbes and bacteria are known to produce organic chemicals from decaying organic matter. Two important organic chemicals that bacteria produce are Ethanol and Ammonia. Ethanol is an alcohol and one of the important organic chemicals listed in our base model and can now be accounted for in the Len Cram model for opal synthesis.

Ammonia is probably the best known of the stober catalysts, an essential opal ingredient, which has been used to produce small size opals, by the Russian team. The Russian method, however is almost sterile, unlike the natural opal environment.

Len Cram's list of opal ingredients is glaringly devoid of any ammonia source, yet we know it is an essential ingredient for opal genesis. Len's method much more approximates nature in that he produces his opal from clay.

In the natural opal environment soils contain materials of decay which are contaminated with bacteria and microbes.

Whilst these bacteria live they excrete organic compounds which provide the right chemical conditions for opal growth, especially by excreting ammonia. The small size of these micro-organisms and their ability for explosive reproductive growth means that they can get to the individual silica spheres where they excrete the ammonia directly to the surface of the silica.

A problem with liquid ammonia is that it is very reactive, and will not persist more than 24-48 hours in solution. The ammonia combines with other elements to form salts, such as chlorides and sulfates. The Sulfates tend to be of low reactivity, as is ammonia in solution unless chloride salts are present.

Bacteria however will persist for very long periods in solution even in some of the harshest conditions. The ammonia supplied by these bacteria will continue for the lifetime of the bacteria, an ideal source of ammonia.

Len's bacteria are sourced from:

- Natural Opal Dirt [clay]
- Mound Spring water
- Acid Sulfate water.

All these environments contain an abundance of bacteria and microbes.

For the massive opal deposits of Lightning Ridge and the Coober Pedy regions, large numbers of Bacteria or microbe colonies, would be needed to supply the quantity of organic chemicals required to facilitate the silica sphere growth that resulted in these massive deposits of opal.

5.19 Nuclearites

It is the Nuclearite that initiates the ion exchange process. Nuclearites also provide a centre for the silica as it condenses out of solution in order to form spheres. This process has been likened to the formation of raindrops in clouds. Some small particles like dust allow the water to come out of the vapour to form rain drops. There are a number of nuclearites that have been identified as being involved in the opal formation process.

Polymorphic Target materials are:

- Bone
- Wood
- Shell
- Possibly even opal clay itself, but it is probable that this clay already contains tiny particles of nuclearite. These

nuclearites reinforce the notion that the opal process is a polymorphic process.

Nuclearites that are identified with opal genesis are listed under the heading "Mound Springs" in Chapter 2. One of the most prominent is Zirconium Hydroxide spheres of small size, in recent studies have been observed filling the gaps between the Silica spheres.

Bacteria have also been suggested as being opal nuclearites. Bacteria could be expected to be present in organic decay materials, such as wood, bone and shell.

5.20 Signs of Opal Forming

After all the mixing is done the mixtures are left to settle. This settling process is actually called sedimentation. Opal spheres grow in solution and then precipitate to the bottom of the container, where they arrange themselves into arrays of cubic packing silica spheres.

You may not notice anything different for the first 3 days after preparing the opal mixtures. However sometime during the first three weeks you should notice small dark structures forming that appear like whiskers beginning to run through the clay mix. These whiskers continue to grow and lighten in colour until they plainly become seams of clear silica gel.

As more time passes this clear gel begins to take on colour.

When about half the ingredients have become opal, [approx 3 months] then Len gently drains the liquid from the jars so as not to disturb the process. The fact that he drains the liquid to allow the reaction to continue is a clear indicator that this is a dehydration polymerisation reaction [condensation]. This then allows for the dehydration process to be completed, as the whole of the clay becomes opal except for some ingredients which become capping stone, as in nature [Ref: 4].

5.21 Opal Dirt

The best ingredients are always going to be sourced from natural deposits; however, it may be possible to substitute the following clay mix for natural opal dirt. *This is as yet unproven but is a possibility.*

Sedimentary Opal in Australia is normally found in a clay mix which is called Opal Dirt and is according to Elizabeth Smith, Typically:

- 75% Kaolin
- 20% Smectite [Montmorillonite]
- 5% Illite [Ref: 2] Elizabeth Smith.

In the book 'The World of Opal' Len Cram gives us the story of the opal miner's cat and a story about fence posts which have become opalized in modern times and the timeframes were 50years or less.

5.22 A More Natural Approach

Len has modified the Russian method by placing it into the clay mix with a special electrolyte that facilitates ion exchange/migration. His opal can be made to nearly any size, which is a vast improvement over the early Russian method.

There are two things that occur in nature and in Len Cram's method that are not present in the early Russian Synthesis method. These are:

- This process is much closer to the natural process of opal formation as normal opal forms within the clay [also called Bulldog shale].

- Clay also contains Alumina which hydrates to form first Aluminium hydroxide which is a silica polymerization activator and then dehydrates to Aluminium Oxide which is a drying agent.
[See Heading: Aluminium hydroxide].

Len Cram also uses Feldspar; this is an addition to the Russian ingredients. What's the reason for using feldspar? The answer to this question lies in the hydrolysis reaction. In water feldspar releases some alkaline ions [Potassium] and some Silicic acid forming Kaolin clay. This hydrolysis reaction releases both more dissolved silica and the silica dissolving agent potassium, and raises the pH of the solution facilitating silica sphere growth.

5.23 Diagenesis [Burial & Pressure]

"The major source of silica for diagenesis is biogenic opal; therefore, silica is especially prevalent in deep marine sediment from active upwelling zones….." [Peter A Scholle, and Dana S Ulmer-Scholle].

Diagenesis is the process by which the burial of organic matter in opal dirt may result in the formation of opal within the sediment or the formation of polymorphic opal, also known as fossilization, or opalization of bones, wood or shell.

What then is the relevance of biogenic opal. It is thought that the precursors for opal formation must be silica of very small particle size.

Some believe that simply dissolving silica will not result in particles small enough to act as an opal precursor, and it is extremely hard to dissolve.

On the other hand, it is known that biogenic opal is the within the size range to become an opal precursor, and is unstable compared to natural silica grains, therefore much easier to dissolve.

Sources of biogenic silica are:

- Sponge spicules
- Diatoms
- Radiolarians

As much of inland Australia was once an inland sea, it becomes clear that there would be large deposits of biogenic silica within the sedimentary layers. As the sedimentary layers are mostly clays of the montmorillonite type, and the temperature range is suitable, all that is then required was organic matter from the decay of dead animals or even vegetation.

5.24 Air Drying of Opal Gel
Len recommends air drying over approx 6 months which still leaves cracking & crazing.

The outcome of this is that the opal produced despite having good colour will not be a marketable product, unless a better drying method is found.

Polymerization produces an Opal Gel that is still soft. It needs to be dried to become the solid opal that we know today. The drying stage is a very vital part of the Opal formation process. If too much water remains within the Opal it will probably crack or craze and the result will be an unstable product.

Len Cram suggests air drying the Opal Gel over a period of approx. 6 months, however the resultant opal contains cracks and crazing. This is not suitable for a salable product. It is probably the result of the opal containing too much water. Even some natural opal has this same problem.

Andamooka Opal is considered to be one of the most stable in the world because of its low water content.

It is quite apparent from this information that a good drying procedure is needed. It should be one that will enhance the opal by producing the correct hardness without the cracks and crazing. Len does not tell us how to do this and his method of air drying is very time consuming yet still allows for cracks and crazing in the finished opal.

5.25 Mineral Oil

Len Cram suggests that mineral oil [white petroleum jelly – Vaseline] may have the ability to heal crazing in opal if it is immersed in it for a long period of time 6-10 years. Better to keep opal in mineral oil than in water. This is due he says to mineral oils ability to act as an ion migration medium.

It is due to this ability of mineral oil that it may be useful in the drying stage of opal gel. The most benefit would probably be gained by using it with the natural method of drying. A small amount added to the clay maybe beneficial, however this method is probably the most efficient at reducing the risk of crazing of all the drying methods.

TABLE 22: Cram Ingredients [Time Approx 12 months]

Ingredients	TEOS
	Water
	Ethanol
	Mound Spring water pH 8.7 plus Acid Sulfate water & TEOS in correct proportion.
Clay	Natural Opal Dirt from opal fields.
	Source of Al & Si ions
Function	
Feldspar	Source of Al, K & Si ions & pH adjuster
Alumina:	Comes from Clay & Feldspar
	Len only suggests that you need to combine two atoms of Alumina to three atoms of Oxygen = Al_2O_3 Aluminium Oxide
Temperature	Not to exceed 60°C (Room Temperature)
Drying	Len suggests air drying over approx 6 months
pH	Adjust Electrolyte requirements or Above 7

5.26 Information resources Len Cram:

1. Creating Opals [Andrew A. Snelling]
http://www.answeringenesis.org/creation/v17/il/opals.asp?vPrint=1

2. Secret of Growing Opals

3. The World of Opals [pgs 22-37]

See also:

Growing Opals – Australian Style

Creation 12(1):10-15, 1989

Andrew Snelling

The Scott Wilson Method [Ref: 15]

5.27 General Information

In the past there have been many attempts to fabricate opal from low cost starting materials such as Sodium Silicate. The result of these endeavors was failure which was thought to be caused by one of two reasons:

1. Sodium Silicate and the resulting Silicic acid were not reactive enough.
2. There was too much variance in particle size of the Silicic acid produced by this method.

However recently [in the last few years] Scott Wilson suggests that Sodium Silicate can be used as a precursor to opal synthesis. This has been made possible by new Ion Exchange resins of the Sulfonic acid variety.

Sulfonic acid forms in nature when Sulfuric acid combines with organics and takes on a dark brown colour.

Scott Wilson's method is now one of the most common approaches to opal synthesis, using Sodium Silicate and ion exchange to form uniform particle sizes of Silicic acid, achieved by using these special ion exchange resins.

The trick to this method in synthesizing opal is to control Temperature, chemistry, motion and ph of the solution producing conditions favouring formation of particles suitable for opal synthesis. This should prevent premature gelling which would result in potch. Normal sedimentation procedures should be followed.

Ion exchange resins that produce small uniform silica spheres are now available as Sulfonic Ion Exchange resins. These resins are available from both Dow Corning and from Rohm & Haas [maybe defunct].

It is now evident through this new method that the first reason for Sodium Silicate being considered as an unsuitable opal precursor was completely wrong. The only thing preventing Sodium Silicate being an opal precursor was the fact that the silica particles were:

- Particle sizes were too large overall.
- Sizes were not uniform

Passing the Sodium Silicate through a Sulfonic Ion Exchange resin, such as Dowex G-26 H, solves these initial problems [Conclusion Summary 3]

Let's take a speculative look at how this synthetic opal synthesis method may work.

Step 1. This is a sol-gel process and due to the precursor and method used, there is no stober reaction in this method. You must first pass the Sodium Silicate solution through the Ion exchange resin however the required amount of water may need to be added to the solution beforehand in order to keep the solution thin. In other words, a silica colloid must be produced, this colloid will have particles that are uniform in size and shape. A thick solution would be expected to gell almost immediately making it unusable. The volume of silica to the water should be around saturation point for silica.

Step 2. An electrolyte needs to be added to encourage particle size growth, probably Ammonia, or one of its salts, however lithium salts may also produce good results. NOTE: Mound spring water could be used to provide bacteria as ammonia source.

Step 3. Adjust viscosity to your desired level by adding higher viscosity liquids such as Glycerin, Polyethylene Glycol [PEG] or a mineral oil. The higher the viscosity, the larger the particle sizes. The larger the particle sizes the rarer, the colours produced [e.g. Oranges & Reds].

Step 4. Sedimentation. Options could include the use of silicone oil or electrophesis to speed up this process.

Step 5. Drying process.

Special Equipment required: Ion exchange column of some type, column or fluid bed. This type of equipment should only be used by experienced operators.

The Bon Ami Method

5.28 General Information

Sometime in the early 1980's a method of creating opal from a silica source and a powder cleanser called Bon Ami or Dutch cleanser was developed. This method was apparently posted in some rock hound magazine and may have become popular for a short time. The method was said to produce good precious opal but there was no way of drying the opal at the time that was produced by this method. This opal was simply kept in glass vases or glass jars under water as a decoration or ornament or even as a curiosity.

However, a method of drying does indeed exist, and this information is available in my expensive business report:

"*The Synthesis of Opal, Step-By-Step*" By Richard Shepherd © 2011. For Further details contact author.

The reference states that Aluminium metal should be used, ground to powder size using a grinder and then heated in suitable cookware for 1 hour. The reason for the heating seems to be to destabilize the metal so that it will breakdown in the electrolyte solution to Al^{3+} ions, much the same way as Alumina Hydrate does. In nature the Alumina is not supplied by metal Aluminium but by alumina from the clay.

The sugar which forms part of the electrolyte is known to react with alkaline solutions such as the bon ami and the baking soda to form aldehyde.

Table 23.The Bon Ami Method

Bon Ami Method	: Substitutes
2 Quarts Water (1.9 Lts)	
1 Tsp Baking Soda (Bi-Carb)	
1 Tsp Salt	
7 Tbsp Sugar	
1 Tbsp Fresh Step Cat litter crystals	1 Tbsp Silica Cat litter crystals/Sodium Silicate
Aluminium (powdered cooked 1hr)	Alumina Hydrate
1 Tbsp Bon Ami Cleanser	Pumice or Feldspar

It has also been assumed that Silica cat litter crystals will all be the same type of silica.

As far as safety goes, this is probably the safest way to produce opal as the ingredients are far less harmful than those used by other methods. This process remains a good experiment for individuals or even for school classes under teacher supervision.

Note: Bon Ami no longer available in Australia. Bon Ami is essentially Pumice (Feldspar) with a little Sodium Carbonate (Washing Soda) or Calcium Carbonate (Lime).

As I have not used this method, I cannot vouch for the accuracy of this information, however, I heard about this method when I was a teenager but lost interest when told there was no drying method. Today, however I know of a number of very good drying methods, which have been described in detail in my reports.

I have known about this process for a long time but I have been unsuccessful in tracking down the original references, despite many hours of research. If you know the source reference for this method I would appreciate the information. My information was obtained from the website e-How.

> I don't have the original references for this method and cannot vouch for it. However if anyone has the original references or the original story on this method I would appreciate a copy. Just send to my email address:

Author Contact Email

RichardShepherd678@yahoo.com.au

The Drying Processes

5.29 Compression Drying

Some manufacturers of synthetic opal use a process called Hydrostatic drying, where they use water pressure as a drying agent. The pressure of the water acts like a mold, squeezing excess water out of the opal gel.

However, this process requires expensive equipment, and takes over 12 months to complete. This is far too long when considering a return on investment.

Taking this into consideration there is a much better process, something more natural, something which we know happens if we really think about the process. It is the compression that occurs as the opal is being formed. As the opal grows within the sediments forming its own seams, through the process of polymerization it is squeezed by the weight of the sediment. As it forms from the clay sediments themselves there is little chance that it is precipitated and fills cracks and voids, this is a myth. There is also the probability that as the particles form from the opal gel, their size increases slightly helping to create compression pressure which acts much like a compression mold. In fact opal is most often found encased in opal dirt or rock which has been baked hard by the heat of the outback [arid environment].

The spent electrolyte will be pushed into the clay which will absorb the liquid. Some of this liquid will of course be evaporated away. The Lipids on the other hand are not so easily evaporated, that is the fats, oils and waxes tend to be incorporated into the clay, preventing the clay from cracking when it dries. As the clay begin to dehydrate, contracting around the opal, it squeezing out excess water from the opal and forces closed cracks and crazing in the opal structure. The clay in which it has formed has become the perfect mold

This present possibility; some of the opal dirt was moist at the time of the opal formation, but not all the opal dirt was moist. Opal only formed within the moist clay. The clay that had not become opal, became the molding material. As the clay baked in the hot outback climate, it dried and hardened, as did the opal within. Here we have a natural molding process by which the drying of opal can be explained. This would also seem to be the most logical explanation as it also explains why opal is found encased in opal dirt.

If you have done any experiments with sun drying clay, especially Kaolin, you will find that it tends to develop large cracks as it dries. These cracks are undesirable as they would almost certainly lead to the opal itself cracking. The importance of a binder which acts like an elastomer becomes glaringly obvious. The Lipids provide this elasticity to the clay, allowing it to dry without cracking. The extra weight of the sediment above would also play an important part in preventing cracking.

Most sources agree that the temperature of the sediment is 50°C or below, this then indicates that supercritical drying is an artificial process and should not be used if one intends to use natural processes for the formation of opal.

Despite the fact that Len Cram believes that crazing is caused by Atom migration it is this authors opinion that the crazing is caused by the drying process as described in the quote above from: Forming Ceramic Green Body and also supported by the following quote:

"Even if the process of drying is carried out very slowly (over more than three months), the capillary forces and the forces of surface tension in the pores between particles leads to the formation of internal stresses – which sooner or later generate a network of 'breaks' (crazing) over the total volume of the opal."

[Ref: 1, Some aspects of Precious Opal Synthesis; pg 3].

In order to produce good quality precious opal this problem must be overcome, how do we do this?

5.30 Opal Molding in Nature

1. Clay; The molding process of the clay not only helps squeeze out excess water but helps to close up the capillary channels created by the escape of this water. It is the capillaries that form the cracks because they leave open channels throughout the opal volume, channels that contain air. Closing these capillary channels is of paramount importance as this will stabilize the opal, preventing crazing, and result in a salable end product. Opal dirt [Clay] should be used as the molding material and it should be sun baked at temperatures high enough to bake the clay, suggest 35-50°C.

2. Ironstone; Volcanic opal becomes encased in ironstone, which acts in much the same way as the clay, providing a molding pressure on the opal thus forcing these capillaries and cracks to close.

Crazing can also occur if opal is left in the sun, this could be due to Atom migration, or it may possible be that drying or dehydration has started up again, opening up capillaries for the moisture to escape.

It is only common sense then to mimic natures way as it produces a very good result, better than any other method that I known to man. Results can be achieved in a short time 1-4 weeks rather than months.

5.31 Aluminium Hydroxide

Al_2 (OH)$_3$; also helps in the drying and hardness of opal, is a natural polymerizing catalyst often employed by industry. Aluminium Hydroxide comes from the clay as the bonds with the silica are broken, the Al3+ ion hydrates: Refer Ch 5. Figure 6.

In the book "The World of Opal" Len Cram states that if you can make two Alumina ions and three oxygen ions to combine and the conditions are right, you will form opal.

This is achieved by the dehydration of Aluminium Hydroxide, as the mixture dehydrates, so too does the Aluminium Hydroxide forming Aluminium Oxide [Al_2O_3].

As already mentioned Aluminium hydroxide is a Silicic acid polymerizing agent whilst the Aluminium oxide is a desiccating [drying] agent.

Related Topic

Artificial Opal; Imitation Opal; Silica; Photonic Crystals; Opal Cracking; Atom migration; Ion exchange; Drying Temperature; molding; sediment;

Capillary forces.

At best opal exists as opal for thousands of years, not millions, as it is part of the silicon cycle and must continue to follow the silicon cycle.

Further Reading

Controlled Growth of Monodispersed Silica Spheres in the Micron Size Range.

By: Stober & Fink

Origins of Precious Opal [Read in light of Len Cram's work]

By: Darragh, P. J., Gaskin, A. J., Terrell, B. C., and Sanders, J. V. (1965) Nature. V209, 13-16.

Gems & Gem Materials

By: Kraus E.H., & Slawson C. B.

McGraw-Hill, NY 1941 [Pages 38-39; 188-91; 252-3].

The Genesis of Natural Opal

Paper
By: John V. Sanders & Peter Darragh [CSIRO]

Other books and magazine articles on opal

Colour of precious opal.

By: Sanders, J. V. (1964)

Nature. V204, 1151-1153.

Diffraction of light by opals.

By: Sanders, J. V. (1968)

Acta Cryst. A24, 427-434.

Microstructure and crystallinity of gem opals.

By: Sanders, J. V., (1975)

Amer. Mineral. V60, 749-757.

The role of aluminum in the structure of Brazilian opals.

By: Bartoli, F., Bittencourt-Rosa, D., Doirisse, M., Meyer, R., Philippy, R., and Samana, J. C. (1990)

Eur. J. Min. V2, 611-169.

A solid state ^{29}Si nuclear magnetic resonance study of opal and other *hydrous silicas.*

By: Adams, S. I., Hawkes, G. E., and Curzon, E. H. (1991)

Am. Min. V76, 1863-1871.

Opal: South Australia's Gemstone

Editors: L.C. Barnes; T.J. Townsend; R.S. Robertson; and D.C. Scott

Published by:

Dept Mines & Energy Geological Survey of South Australia 1992

Biosilica formation in diatoms: Characterization of native silaffin-2 and its role in silica morphogenesis.

By: Nicole Poulsen, Manfred Sumper and Nils Kröger

Lehrstuhl Biochemie I, Universitätsstrasse 31, Universität Regensburg, 93053 Regensburg, Germany

http://www.pnas.org/cgi/content/full/100/21/12075

Silicatein: Cathepsin L-like protein in sponge biosilica.

By: Katsuhiko Shimizu, Jennifer Cha, Galen D. Stucky, and Daniel E. Morse.

http://www.pnas.org/cgi/content/full/95/11/6234

Vol 95, Issue 11, 6234-6238, May 26, 1998

Mechanism of Growth of Supramolecular Crystals in Concentrated Suspensions of Mondispersed Spherical Silica Particles (MSSP)

A.F., Danilyuk; V.V., Serdobintseva; D.V., Kalinin

International conference of Silica Science & Technology SILICA-1998, Mulhouse, France, Sept. 1998.

Opalization of Fossil Bone and Wood: Clues to the Formation of Precious Opal [Pages 264 – 268]

By: Benjamath Pewkliang [1,2], Allan Pring [2] & Joel Brugger [1,2]

[1]CRC LEME, School of Earth and Environmental Science, University of Adelaide, SA 5005

[2]Department of Mineralogy, South Australian Museum, North Terrace, Adelaide 5000, South Australia

Regolith 2004

Related Websites:

http://www.answersingenesis.org/creation/v12/i1/opals.asp

http://www.austgem.gil.com.au/opalsynth.html

http://www.opalsdownunder.com.au/articles/fields.htm

http://www.colonialopal.com.au/types.html

http://www.costellos.com.au/opals/valuing.html

http://www.opalhut.com.au/opal.htm

Weathering of rocks

http://fbe.uwe.ac.uk/public/geocal/soilmech/classification/soilclas.htm

#CLASSWEATHER

Opal related Patent

US Patent # 3,497,367

Opaline Materials and Method of Preparation

By Sanders & Darragh

Other Related Patents:

US Patent # 3,325,321 Shannon, June 1967

"   "   # 2,574,902 Bechtold November, 1951

"   "   # 3,301,635 Berga January, 1967

German Patent

1,111,775 Germany, July 1955

Fire Opal: Inventor: Bhandari, Rajneesh of Jaipur, INDIA.

US Patent No. 7465421.

NOTE: The patent [US # 3,497,367] developed by Sanders & Darragh of the CSIRO was to protect the Opal industry from man made opal. This patent is based on old information and is not relevant to the newer methods of Len Cram, or Richard Shepherd. Some of the heat treatments mentioned in this patent are totally unnatural and not used by these newer methods. This patent is based on the use of TEOS which does not need to be used to make opal. Dissolving silica [clay] with Carbonic acid is not difficult and avoids the use of TEOS altogether. The real trick to making opal is growing the silica spheres and hardening of the opal. I have provided you with methods to do both.

It remains your responsibility to avoid patent infringement.

Modifying existing methods is often a way around infringements but you need to be able to show that your method is an improvement over the existing method.

Bibliography:Chapter 5

Opal Synthesis
1. *Some Aspects of Precious Opal Synthesis* [Pages 1-5]

S.V. Filin, A.I. Puzynin, V.N. Samoilov

Scientific Centre for Applied Research, Dubna, JINR, Russia

http://www.austgem.gil.com.au/opalsynth.html

[Report] Current at 15/05/2003

2. *Black Opal fossils of Lightning Ridge*;

by Elizabeth Smith; Published   by Kangaroo Press 1999).
[Book]

3. *Growing Opal Australian Style*

Snelling A.

Creation [Magazine] 12(1):10-15, 1989

4. *The World of Opals*   [Book]

Author: Allen W. Eckert
Publisher: Wiley 1997   Recommended Reading

[CH 2 pgs 17 – Ch 3 pg 22- 45]

5. *The Chemistry of Silica, (Solubility, polymerization, colloid...*

Published by Wiley, New York c1979
    Author: Iler, Ralph., K   [Book]

6. *How is Opal formed? The Geology of Opal*

Website Article

http://www.opalsdownunder.com.au/articles/formed.htm

7. *McGraw-Hill Ency. Science & Technology*
Subject: EDTA
Vol 6, pgs 571
Vol 3, pgs 541
Vol 18 pgs 454

8. *Sedimentary Rock-Hosted Opal   Q08*

by: S. Paradis[1], J. Townsend[2] and G J. Simandl[3]

[1] Geological Survey of Canada, Pacific Geoscience Centre, Sidney, British Columbia, Canada
[2] South Australia Department of Mines and Energy
[3] B.C. Geological Survey, Victoria, British Columbia, Canada

9. *Silicification and the conversion of Sinter to Chert*

University of Aberdeen

http://www.abdn.ac.uk/rhynie/sinter.htm

10. *Formation of Silica Structures Utilizing A Cationically Charged Synthetic Polymer*

(Abstracts 2001 Academic Poster Session, Cleveland, Ohio).

Silicification & Biosilicification
    Parts 2; 5 & 6
By Siddharth V. Patwardhan and Stephen J. Clarson

Http://www.eng.uc.edu/~sspatwar/Abstracts.htm   [Pages 1-4]

11. *Introduction to Modern Colloidal Science*

Author: Robert J., Hunter.

Publisher: Oxford University Press, 1993

12. *Species-specific polyamines from diatoms control silica morphology*

By: Nils Kröger, Rainer Deutzmann, Christian Bergsdorf, and Manfred Sumper

http://www.pnas.org/cgi/content/full/97/26/14133

13. *Infracrystallization in monodisperse system of amorphous silica as a model of infracrystallization and growth of "photonic crystals".*

By: V.V. Serdobintseva; ICCG-13 Quoto Japan, June 2001.

http://www.uiggm.nsc.ru/opal/links

14. *Catalytic Power*

[http://www-
biol.paisley.ac.uk/Courses/StFunMac/glossary/catalytic.htm

15. *Scott Wilson Opal Synthesis 31/01/2005*

The New Mexico Facetor [Pgs 1-3]

By Nancy L. Attaway & Scott Wilson

http://www.attawaygems.com/NMFG/Program_speaker_sco
tt_wilson_opal.html

16. *Decomposition*

http://www.socgenmicrobiol.org.uk/pubs/micro_today/pdf/
110108.pdf

17. *Forming Ceramic Green Body*
Wet Forming of Ceramic Powders
Sol-Gel Processing of Ceramics (Polymerization)

Gelation Mechanism involve destruction of the double layer
that keeps particles apart & can include: (A & B only)

Http://www.mmat.ubc.ca/courses/mmat382/sections/cnc43
10.doc

18. *The Claisen Condensation*

Brooks/Cole Publishing Company

Website article

http://www.chem.uic.edu/web1/OCOL-

II/WIN/CH23/F2.HTM

19. *The Fate of Silicate Materials in a Peat Bog*

Website citation
Http://www.osti.gov/energycitations/product_biblio.jsp_id=5
539161

20. *In Situ solid state* NMR *studies of Ethanol photocatalysis
characterization of surface sites and their reactivities*

By: Son-Jong Hwang, Daniel Raftery
Publisher: Elsevier
Catalysis Today 49 (1999) 353 – 361

21. *Biogenic Silica Patterning: Simple Chemistry or Subtle Biology?*

MIMIREVIEWS

P.J. Lopez & T. Coradin

ChemBioChem 2003, 3, 1-9

Wiley – VCH

22. *Synthetic opals made by the Langmuir –Blodgett method*

Thin Solid Films 437 (2003) pgs 276-279

Authors: M. Bardosova, P. Hodge, L. Pach, M.E. Pemble, V.
Smatko, R.H. Tredgold, & D. Whitehead

Publisher: ELSEVIER

www.sciencedirect.com

23. Thesis

[http://users.mrl.uiuc.edu/floren/thesis/chapter_3.pdf]

Photonic Crystals based on Silica microspheres [2003]

CH 3. Artificial opal fabrication methods [pgs 72-87].

By Florencio Garcia Santamaria

University of Illinois at Urbana-Champaign, USA

24. *Synthetic or Gilson black and White Opals*

http://www.opalsinformation.com

1 Sept 09

25. *Glass Opal*

Subject: Slocum Stone

http://www.emporia.edu/earthsci/amber/go340/create.htm

1 Sept 09

26. *Slocum stone is a recent (1976) opal imitation named after its Canadian inventor.*

http://www.aigsthailand.com

1 Sept 09

27. *A Colour Guide to the Petrography of Carbonate Rocks: grains....*

American Association of Petroleum Geologists, Vol 77, pg 408

Nature 2003

Authors: Peter A. Scholle & Dana S. Ulmer-Scholle

6 TRICKS & TREATMENTS

6.0 Sedimentation Process

The common method of sedimentation called "Gravitational Sedimentation" is very slow taking 6-9 months. It is therefore advantageous to find ways to accelerate this important process.

Two ways are:

Firstly by adding Silicone Oil, and

Secondly by electrophoresis.

6.1 Silicon Oil [Ref: 6]

Scott Wilson also suggests adding 10% methanol to the ethanol possibly to help the drying process. He also suggests that sedimentation be done under silicone oil, citing Philipse 1989. The solution needs to be adjusted to match the viscosity of the oil. The sedimentation vessel should be covered to keep out dirt and dust. The Silicone oil is placed at the bottom of the sedimentation vessel and the solution is poured on top of the oil. As the solution loses ethanol and water by evaporation through the oil, the opal gel sinks to the bottom of the oil as a blob. Colour improves greatly as the blob shrinks.

6.2. Electrophoresis

There is a method that can be used most effectively which can halve the settling time, it is called electrophoresis. Basically this process increases the sedimentation velocity. Electric fields are used to drive the sedimentation velocity of to 0.4 mm per hour which is approx. double the natural rate. A settling period of 7 months could be cut to 3.5 months, making this method very attractive to anyone thinking about synthesizing opal [ref: 5].

Electrophoresis also improves the reliability of the settling process, as sometimes without it the settling process can fail altogether.

NOTE: There is no mention of ammonia being used here. Ammonia is a catalyst but the process can probably proceed slowly without it. However in industry Acid or Alkaline catalysts are most often used.

See Online: www.Sol-Gel Gateway.com

www.psrc.usm.edu/mauritz/solgel.html.

6.3 Alternative Electrolyte [Ref: 4]

From Infracrystal research (Photonic Crystals)

Electrolyte (disperse medium):

• Water, Ethanol, Ester (fats, oils & waxes), Acetone
• Require counter ions; NH_4^+, K^+, Na^+.

Note that Esters, Fats, Oils & Waxes are higher viscosity liquids.

6.4 Drying Opal Gel -- Methanol

Methanol is yet another product of organic decay. Methane gas is produced first but if the gas bubbles through water methanol or wood alcohol forms. Scott Wilson claims that methanol sharply decreases the drying time of opal gel. He suggests adding 10% by vol to the ethanol for maximum results. Keep in mind methanol is very toxic and can enter through your skin. Appropriate gloves should be worn when using, take extreme care.

6.5 Manipulating Colour

There are a number of tricks and treatments for the manipulation of colour in opal. The most important of these are:

- Impregnation with various, oils, waxes or plastics. These are used primarily to disguise fractures or crazing and improve play-of-colour.
- Smoke is sometimes used to produce a black opal.
- Treatment with dyes [aniline] or chemicals to make light opal black or to improve the play-of-colour.
- Carbon or sugar treatment is used in an attempt to darken the opal to give it a black body colour. Sugar and Sulfuric acid baths are sometimes used which result in a darker colour with a pin fire effect.
- Silver Nitrate is used to blacken some opals, in a process similar to the one used to blacken pearls. The opals are soaked in silver nitrate solution. A light is then shone on them, causing the precipitation of metallic silver, rendering the body colour black.

All these methods however, are detectable by a professional although they may escape the attention of an amateur. Nearly all these methods result in the lowering of the SG and the stones may become more porous and therefore less stable.

6.6 Sonification

Ultrasound is often used to ensure that silica particles or spheres remain separated, as they normally do have a tendency to aggregate. Sonification is one of the few ways to prevent silica particles bonding. In opal synthesis there is a need to keep the particles separated whilst the spheres are developing, hence the term monodispersed particles. Thus meaning single dispersed or separated particles.

6.7 Centrifuging

Centrifugation is sometimes used to purify solutions or to separate out particles of the same size from the solution.

6.8 Solution Viscosity and Sphere Sizes

The viscosity of the solution can be affected by dissolved matter, the most significant being salts and clay content. The solution containing all the dissolved matter is sometimes referred to as the Supernat or the Mother Liquor, or simply as the electrolyte.

Two important features of the solution are the viscosity of the solution and the conductance of the solution. Viscosity is a measure of the density of the liquid.

Neutral chlorides are known as leaching salts due to their ability to dissolve metals from rock, soils and even clays, keeping them in suspension which raises the conductance of the solution and thickening the solution at the same time. These chlorides could leach the Alumina from the clay releasing the Acidic Al_3+ ion which then joins with the Acid Sulfate salts. The leaching of Alumina from the clay allows the silica to dissolve much more easily. The Silica Alumina bond is a very strong bond which causes the silica in the clay to be extremely hard to dissolve. Yet these salts have provided a way to break this bond and to separate these two ions allowing the silica to dissolve and to form silica spheres which are essential for opal genesis.

One of the keys to obtaining the larger sphere sizes, which produce the rarer colours is to use a *higher viscosity supernat*. Some of these chemicals are Ethylene Glycol produces particle sizes over 550nm. Silaffins 1 & 2 can produce silica spheres between 100 – 1000nms, which means that LCPA's and PEP as substitutes are capable of the same.

Amino acids containing Nitrogen increase sphere size.

Amino acids containing Hydroxyl groups or hydrophobic groups produce smaller sphere sizes.

Viscosity is also affected by the addition of thicker liquids such as; Fats; oils; waxes; gums; glycerin; and resins which are naturally produced of decay. These can all be dissolved by Ethyl Acetate which can form as an intermediate product of Ethanol as it oxidizes to form Acetic Acid. A question to be raised here is, "Does Ethyl Acetate dissolve Silica".

6.9 Conductance of Solution

The electrolytic properties of the solution are greatly enhanced by the addition of salts and metal ions in the solution. These metal ions not only increase conductance but also increase the viscosity of the solution by increasing its density.

Chloride salts and Acid Sulfates dissolve metals from surrounding rocks, soils and clays through which they flow, suspending these metal ions in solution. If ammonia is also released, it too will trap metal ions in solution by the process known as solvation. Thus a very strong electrolyte is built up by these processes.

6.10 Double Dissolution of Salts

A common element found in all the opal environments seems to be Acid Sulfates and neutral Chlorides. Research may indicate that Acid Sulfates are the result of the hydrolysis of FeS_2 producing Sulfuric Acid and the Fe_2^+ and Fe_3^+ ions, however, they can also contain any of the Acidic ions; Al_3^+, NH_4, Pb_2^+, Sn_2^+ and even Phosphoric Acid. Neutral Chlorides likewise are not just NaCl but may contain any of the Neutral Cations, but the common cations in these brine salts are Na^+; K^+; Mg_2^+ and Ca_2^+ [see: CH. 2].

These two salt species have the important ability of being able to dissolve one another. The Acid Sulfates dissolve the Chlorides which then become acidic, the acid sulfates revert to sulfate salts. They achieve this by swapping Hydrogen ions between them. This acid swing continues until total neutralization of the acids leaves both elements as salts. This process must play the ultimate role in converting soluble silica into opal spheres.

6.11 Lithium Salts

As an alternative to Ammonia there is the Lithium salts, Lithium Chloride would probably be one of the best. Although lithium salts would be rare in nature, they are used in industry, specifically in PEO, for producing Silica spheres. Lithium shares two of the most significant properties of ammonia, these being:

- An excellent electrolyte, used by industry in batteries and
- Produce solvated electrons

Due to these properties, they can be used as a substitute to ammonia in the synthesis of opal, but would probably not be involved in natural genesis of opal.

Lithium salts are toxic.

6.12 The Role of salts [Ref 3]

The major functions of salts in the opal genesis process are:

- Break the Al-Si bonds allowing silica to dissolve more readily.
- Salts such as the Chlorides break the Si-Si double bond allowing the silica to dissolve into monodispersed particles.
- Act as gelling agents
- Increase conductance of electrolyte by being salts, and by the metals added to the electrolyte due to leaching ability of salts.
- Increase the viscosity of the solution by increasing density both by salts and by added metals.
- Acid Sulfates catalysis of organic decay products, releasing organic chemicals into the mix. Some of these organic chemicals dissolve silica.

Excessive salts result in premature gelling, interfering with particle growth, resulting in potch [common; worthless opal] at best.

In the opal environment Sulfates and chlorides are found typically at a concentration of up to 3%. This concentration then should be used as the standard to begin with, and then it can be varied up or down according as trial and error to find the optimum concentration. It would not be unreasonable to think that this 3% salt solution must be precipitated for the formation of opal to occur uninhibited. 3gms per 100gms [or mls] water equals 3%. Normal seawater contains 35mls NaCl per 1000gms, which is 3.5 gms per 100gms or 3.5%.

Salts can be precipitated by using small quantities of organic solvents such as, Ethanol, Acetone, or Isopropanol [Rubbing alcohol]. In this way salt density can be controlled in solution. Thus bacteria which produce ethanol may also serve in the role of regulating salt density around the silica spheres. Excess salt falls to the bottom where it may assist in gelling the precipitated silica spheres. Salts help the gelling process.

6.13 Accelerating the Process
- *Sedimentation*

Normal sedimentation often called gravitational sedimentation which includes the settling of silica spheres [aggregation & precipitation] and the self-assembly and alignment which forms the opal gel, or opal cake. The time range for this process can vary from 6 - 9 months which is a considerable amount of time.

-*Electrophoresis*

Sedimentation times can be cut in half by using a method called electropheresis. This is a great time saving, which could cut the cost of production by having the product ready up to 3 months early. See description of electropheresis in previous chapter under the Russian Method.

- *Silicone Oil*

Using silicone oil is also believed to speed up the process; the oil forms a film layer on the surface of the supernat. It is thought to accelerate the dehydration process.

- *Drying the Opal Cake*

Drying the Opal gel has been one of the biggest problems associated with making synthetic opal in the past, it has been time consuming, with less than perfect results. However nature has given us a fast efficient drying method that produces great results. See The Shepherd Method of drying.

Bibliography:Chapter 6

1. *Microstructure and mechanical properties of synthetic opal: A chemically bonded ceramic*
Authors: Thomas C. Simonton, Rustum Roy, Sridhar Komarneni, Else Breval.
[pgs 1-4]

2. *Gems made by Man*

By Kurt Nassau

3. *Mineral formation and redox-sensitive trace elements in a near-surface hydrothermal alteration system*

Source:
http://www.osti.gov/energycitations/product.biblio.jsp?osti_id=687712

Authors: Gehring, A.U.[ETH Zurich, Schlieren (Switzerland). Inst. for Terrestrial Ecology] | [ETH Zentrum, Zurich (Switzerland). Office of Planning]; Schosseler, P.M.; Weidler, P.G.[ETH Zurich (Switzerland). Inst. for Physical Chemistry].

4. *Infracrystallization in monodisperse system of amorphous silica as a model of infracrystallization and growth of "photonic crystals".*

By: V.V. Serdobintseva; ICCG-13 Quoto Japan, June 2001.

http://www.uiggm.nsc.ru/opal/links

5. Thesis

Photonic Crystals based on Silica microspheres [2003]

CH 3. Artificial opal fabrication methods [pgs 72-87].

By Florencio Garcia Santamaria

University of Illinois at Urbana-Champaign, USA

http://users.mrl.uiuc.edu/floren/thesis/chapter_3.pdf

6. *Scott Wilson Opal Synthesis 31/01/2005*

The New Mexico Facetor [Pgs 1-3]

Subject: Silicon oil.

By Nancy L. Attaway & Scott Wilson

http://www.attawaygems.com/NMFG/Program_speaker_scott_wilson_opal.html

7. *Opals from Sand*
http://www.russia-ic.com/education_science/science/breakthrough/665/

RICHARD SHEPHERD

7 CONCLUSION

7.0 Basis Processes

When investigating sedimentary opal formation it should be
noted that there are actually two slightly different methods
used in nature. The first is opal that forms in clay and creates
its own seams, it grows through moist clay by a process called
polymerization. The second opal process is called
polymorphism, as the opal grows within the structures of
wood, shell or bone, taking on the shape of these structures, as
if molded. These two methods are different and will be treated
separately.

7.1 Zeroing in on the secrets of Opal Genesis

The search for the secrets of opal genesis, or a mystery
opalizing fluid, has been investigated by many researchers
without clear results. My conclusions however have
established a much clearer understanding of the natural
formation of opal. My conclusions have been found by
comparing different methods, and opal formation models, and
by filling in some gaps by using deductive reasoning. The
following methods and reactions in particular were studied,

compared and used to arrive at these conclusions:

Base Model; The Russian Method

- The Len Cram Model
- The Microbial Model
- The Byron Deveson Model [Mound Springs].
- Colloidal Science/The Sol-Gel Method
- The Stober Reaction
- Environmental/Natural: Acid Sulfate and neutral chloride mixing.

7.2 The Base Model

The Base model provided us with pivotal information about both the ingredients needed and the method required to produce synthetic opal.

The Base model is used like a control in a science experiment. It gives us a list of required ingredients, the chemical reactions involved, and outlines a method that works to produce synthetic opal. It is produced in an almost sterile environment unlike natural opal.

We can reasonably expect that natural opal will form in a similar way, having a requirement for the same, or similar ingredients.

7.3 The Len Cram Model

Len Cram's model differs to the base model in that the opal is produced in clay. In comparison to the base model it became obvious that Len has not mentioned using the organic chemicals, Ammonia or Ethanol. Both these chemicals are required for creating opal.

Now, the question is raised, "Where does Len get his organic chemicals for his opal to form"?

7.4 The Microbial Model

The Microbial Model states that bacteria or microbes are essential to the natural opal forming process. When you look at the opal environment in nature, it becomes hard to argue against this view. Bacteria are present in nearly all soils, and opal clays provide an excellent habitat for microbes.

Microbes are present in the air, they are present in water, and they are present in decaying organic matter. In fact the Great Artesian Basin is an excellent source of microbes and bacteria.

The Australian opal fields are almost entirely within this region.

Microbes are known to accelerate the decay of organic matter breaking it down into organic chemicals, two of the most commonly produced organic chemicals are Ammonia and Ethanol. Microbes are small enough to invade colloidal silica and to excrete these chemicals directly onto the surface of the silica spheres. As long as these microbes live they will continue to excrete these chemicals. This provides a long lasting supply of these organic chemicals to the silica spheres.

7.5 The Mound Spring Model

Some people think that large amounts of pressure are required to produce opal. Byron Deveson seems to be one of these people. However neither the Russian Model nor the Len Cram model require this type of pressure.

Deveson however does explain in some detail the role of clay in opal formation. Montmorillonite has special ion exchange properties, and comprises a large proportion of opal dirt. This clay is essential to sedimentary opal formation.

7.6 The Mystery Opalizing Fluid

Byron Deveson was also correct about Mound spring water being part of the opalizing fluid. Many of the mound springs around the GAB have the right chemical composition to produce opal. Mound spring waters fed by the GAB contain a rich source of dissolved silica and microbes, which flourish in these waters.

The Russians gave us a list of essential ingredients for opal creation, this can be regarded as the bench test for opal ingredients. The minimum requirements, which are the essential ingredients. If any system claims to lack these ingredients, we need to look at other ways that these ingredients are supplied to the specific method in question.

This is where Len Cram's source of Ammonia and Ethanol are found.

He uses mound spring waters, adjusts the pH using some acid sulfate waters, which also contain microbes, he produces colloidal silica by mixing with Ethyl Silicate [TEOS]. This solution he calls his electrolyte.

As a colloid, this solution remains stable, and has a long shelf life. This is pot number one. Len uses a two pot process. The second pot contains opal dirt with feldspar in it. The feldspar is used to raise the pH so that when the two pots are mixed, it raises the pH above 7, at which point the silica spheres begin to grow. This is the beginning of opal formation.

This is supported by Ilers diagram on polymerization of silica and by the figure at the end of the mound spring information in chapter 2. The information on the mixing of Acid sulfate and neutral chloride waters also supports this idea that opal forming solutions begin acidic and then turn alkaline for the formation and growth of silica spheres.

The fact that Len Crams opal process works is the proof that was required to proclaim the Microbe model as correct. All three models have merit but opal would not form without bacterial involvement.

Figure 19. GAB and Mound Spring Waters

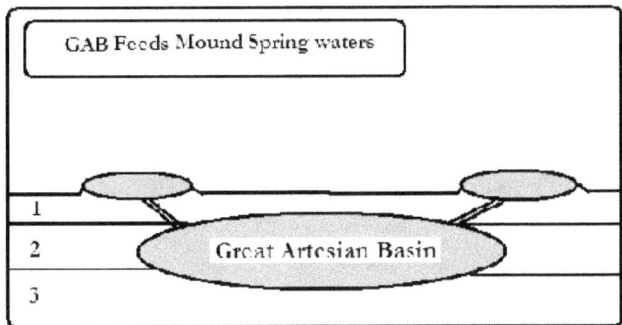

GAB Feeds Mound Spring waters

1

2   Great Artesian Basin

3

1 is Sedimentary Opal Clays
2 is Sandstone
3 Iron Sulfide FeS2

Colloidal Science and the Stober Reaction have already been adequately covered in previous chapters.

7.7 Environmental/Natural: Acid Sulfate and neutral chloride mixing

Studies of the Australian Opal fields has yielded solid evidence of how sedimentary opal formed in Australia. The following quotes bring this evidence together revealing conclusively the manner in which this opal in the Great Artesian Basin and other parts of Australia formed and possibly also in the Virgin Valley in the USA.

> "For the formation of precious opal Darragh et al. (352) point to three features that are important, at least in the Australian fields:(a
>
> An abundant supply of readily soluble silica: (b) an arid climate restricting shallow ground waters to sharply defined dams such as

bentonite beds, which prevent the formation from drying out, thus retaining a typical solution containing up to 3% soluble sulfates and chlorides and 80ppm soluble silica: and (c) the presence of cavities, formed in a variety of ways, in which silica particles can collect and arrange themselves (Ref: Ralph K. Iller 399)".

Here are some other supporting quotes:

"There has been much debate in recent times concerning the composition of Opalising fluids in the formation of opal in Australia. In most other nations the opal is of a volcanic origin and obviously form in volcanic fluids. It is known that the hydrothermal alteration of felsic volcanic by acid sulfate fluids and from the boiling and/or mixing of neutral chloride geothermal brines in epithermal settings (Corbett & Leach 1998, Regolith 2004, pg 264)".

You simply cannot ignore the similarities here, the fact that one is sedimentary and the other volcanic is of little real significance. It is the similarities that are important. The mixing of the Acid Sulfates with the Neutral Chlorides.

"Mixing warm, alkaline mound spring water with cool, slightly acidic ground water with low total dissolved salt content would decrease pH, lower the temperature, and lower the ionic strength of the mound spring water. All three changes favour the formation of Silica Spheres (Byron Deveson, pg 5)".

The important part of these quotes is the mixing of Acidic solutions with the alkaline waters. The most abundant acidic waters are the Acid Sulfates, and the most abundant of the alkaline spring waters are chloride brines.

Acid Sulfate/Chloride waters exist in areas of the GAB and tend to be recognizable by the green colour due to the dissolved iron content.

7.8 When Acid Sulfates mix with Neutral Chlorides

When these two fluids mix a process called "the double dissolution of salts" occurs. This happens as hydrogen ions are swapped between salts.

The expected result of this reaction would be a changing of pH as one salt changes to an acid, this reaction seesaws back and forth until there is a total neutralization of the acid and both solutions revert to salts.

This seesaw reaction causes a corresponding seesaw change in pH allowing for the maximum number of silica spheres to be formed out of the solution.

In comparing the mechanism of opal formation between volcanic and sedimentary it seems that they are very similar both relying upon the mixing of Acid Sulfates and neutral chlorides.

The diagram below represents a summary of the major stages of sedimentary opal genesis in a mound spring environment.

Figure 20. Major Stages in Natural Formation of Opal, Mound Spring Model

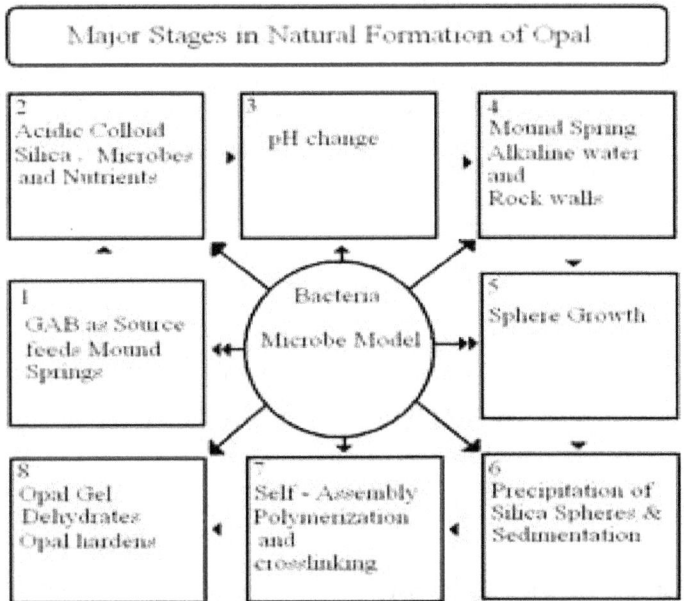

Square No 1. The GAB produces the temperature and pressure to dissolve silica. No 2. shows the ingredients that are injected into the Mound springs, as they travel through sandstone and alkaline rocks and then mix with Mound Spring waters which are Alkaline produces a change in the pH of the mix. Acid to Alkaline. The temperature of the mound springs waters are approx __C. The Alkaline nature of the Mound Spring waters and temperature along with the chemicals being applied to the surface of the silica spheres by the microbes found naturally in these waters produce optimal conditions for silica sphere growth. 6 shows that after sphere growth, the silica spheres then precipitate to the bottom of the mound spring and begin the process of settling on a bed of montmorillonite clay. 7 Indicates the processes of Self assemblage and as the silica spheres form networks polymerization with crosslinking results.

426

8 This polymerization produces soft opal gel [opal cake]. Dehydration of the formation also dries and hardens the opal gel resulting in hard opal.

Let's explore each stage of the above diagram, this will give a clearer understanding of how opal forms in a natural sedimentary setting.

7.9 Opal formation stage by stage

Stage one shows mound springs and their link to the GAB. It is in the waters of these mound springs that opal begins to form. The diagram does not show the breakdown of organic matter, which is covered separately. The presence of DOM in these waters is assumed. The silica that begins the opal reaction is sourced from DOM. This is Biogenic silica which is present in almost all living things.

Spring waters contain bacteria, iron, and some contain sulfuric or other organic acids, which all help breakdown the DOM. Proteins are broken down to release amino acids, which in turn breakdown to release Biogenic silica, Amines [Ammonia salts] and Carboxylic acids. The Biogenic silica dissolves in theses waters.

Stage Three & Four

In stages one and two the biogenic silica exists as dissolved silica. A change in pH is required for the silica to begin to separate out of the water, this is condensation.

Stage four simply explains some of the factors which may cause a change of pH to occur within the mound spring.

Stage Five & Six

The presence of nuclei in the mound spring waters allows the silica which is condensing out of the water to accumulate around these nuclei. They act as centers for the growth of silica spheres. Growth occurs by a gradual build up of layers of silica, this is a lamellar process.

The Silica spheres will remain in solution and continue to grow so long as they remain buoyant. Therefore the denser the solution the longer the Silica spheres remain buoyant and the larger they become. The larger spheres produce the rarer colours. At some point the Silica spheres become too heavy to remain in solution and will fall, or precipitate, to the bottom of the mound spring.

Stages Six and Seven

The Silica spheres land in the bottom of the mound spring which is mainly composed of Montmorillonite clay. Here the spheres begin to self assemble into lines, or strands due to electrical charges of the spheres. This is the beginning of polymerization.

However, due to the bacteria in the clay, the water and around the silica spheres, ammonia from the bacteria accelerates the hydrolysis of the clay which release more soluble silica and accelerates the formation of Aluminium Hydroxide. The Aluminium Hydroxide is a polymerization catalyst for silica, helping in the opal polymerization process. The freed silica is drawn into the opal forming process.

Any ion exchange probably occurs before gel formation, as this is not a soft gel by gel standard.

The silica strands increase in numbers, becoming bundles, these bundles then cross link forming a proper gel. This is opal gel, soft by opal standards.

Stage Eight

In nature opal gel hardens within the clay, it is highly probable that the clay acts to pressurize the opal gel squeezing out excess water. The opal formation dries as the clay dries. This may happen as an entire spring dries out, or a part of the spring dries out. It is a dehydration process caused by the hot natural climate of the area of the Great Artesian Basin.

7.10 In-Situ Water Sources

- Clay Hydrolysis
- Acid & Alkali neutralization reactions
- Protein decay

These water sources are totally independent of any groundwater or outside sources. As protein begins to breakdown between 60- 90% will be released as water. This water hydrates the surrounding clays which then undergo hydrolysis, releasing excess soluble silica and more water. Acids which are released from the stomach may be neutralized by alkaline products, possibly amines, releasing water as a by-product.

7.11 Soluble Silica Sources

- Clay Hydrolysis produces excess soluble silica as stated above.
- Biogenic Silica, abundant in organic matter [All living things contain Biogenic Silica].

7.12 TEOS formation

Ingredients for TEOS formation are as follows:

- Soluble Silica from clay.
- Ethanol from decay.
- Catalyst [either $MMOH_{red}$, or soil oxide catalysts].
- Naturally occurring i.e. Silicon Catecholates

Not all opal formed by this method is polymorphic, so why does opal sometimes form in bones, wood or shell. Yet at other times simply forms large seams or shelves independent of polymorphic materials.

7.13 Nucleating Agents

The answer to the question above is that the presence or absence of nucleating agents determines where opal forms. Nucleating agents draw the dissolved silica like a magnet. They act like seed crystals around which silica spheres condense. They are essential for opal formation.

Nucleating agents are the centers around which silica spheres form. It seems that growth of silica spheres is caused by a build up of layers around this centre, rather than by Ostwald ripening. This build up or layering could be described as a lamellar or laminating process. Layers of silica build up like shells, one on top of the other.

As the spheres become too heavy they fall to the sediment at the bottom of the mound spring. The size of the spheres would be uniform this way. The buoyancy of the solution will determine the sphere size, the longer the spheres stay in solution, the larger they will become, as layer builds on layer [This maybe Ostwald Ripening.]. The viscosity of the solution, as a measure of buoyancy determines sphere size. The higher the buoyancy, the larger the spheres, and the rarer the colours that are produced. [See: Viscosity 4.95]

Lower buoyancy produces smaller spheres and the more common colours. There are a number of factors which affect the viscosity of these waters and therefore the buoyancy of the water. Some of these factors are:

- Salt content
- pH of the water
- Temperature

A change in any one of these factors may cause precipitation of the spheres.

Most of the nucleating agents for opal are listed under the heading of Mound Springs, including Zirconium Hydroxide, it is not yet clear however mounting evidence points to microbes such as Bacteria as a nucleating agent for opal formation. [See: 5:17]

7.14 Pressure

One of the greatest pressures in the opal forming process is probably caused by the expansion of silica spheres in a confined space, and the second greatest by the drying of the encasing clay which shrinks around the opal forcing out excess water and closing cracks and crazing by exerting a force similar to a mold. Huge pressures are also found within the waters of mound springs. [See Mound Springs].

7.15 The Ikaite Question

Does Ikaite play a role in polymorphism? Only further research can answer this question. Ikaite formation requires frigid temperatures, below 10C. It is not known if opal can form at such a low temperature.

7.16 pH Change is Essential

The question of pH change was brought up in chapter three. There is, however, little evidence that a pH change is necessary in opal formation.

Neither the Russian Synthesis method nor Len Cram mention a need to change the pH. Therefore the pH is preset by the formula used, that is to say by the pH of the electrolyte. PH change is required but it is built into the ingredients and method used.

The solution must start as acidic solution in order to dissolve enough silica to saturate the solution but for silica spheres to form and then to grow the pH must change to alkaline.

The reaction between Acid sulfates and neutral chlorides would set its own pH level. This reaction has been examined in chapter three.

7.17 Temperature

Temperature range is extremely important if using microbes or yeasts, which are used by nature. The microbes and yeasts require temperatures between the ranges of 37-50°C. Constant temperatures within this range will provide ideal conditions for microbes to flourish. As already stated microbes provide the Ethanol and Ammonia for the genesis of opal in the natural

432

environment.

7.18 Drying agents
The importance of Aluminium Oxide [Al_2O_3] that forms during the opal formation process cannot be understated in its significance in helping to dehydrate the opal gel to hardened opal.

Scott Wilson suggests Methanol as an opal drying agent also

7.19 Drying Timeframe
Use the natural method for drying; in order to establish timeframe for this process, I suggest you do a time trial. First trial: 1 week drying in the clay; second trial: 2 weeks drying in the clay; Third trial: 1 month drying in the clay. Sun dry or oven dry to 50°C. NOTE some $CaCO_3$ may need to be added to the clay as a binding agent to prevent clay from cracking, if the clay cracks it may also crack opal. If you have a look at opal dirt it rarely cracks but much of it does contain some $CaCO_3$.

The real secret to making Precious opal is knowing how to dry the opal gel. This knowledge is IMPERATIVE. Lack of knowledge about drying the opal gel will only result in failure. Cracked or crazed opal is worthless, in stark contrast to some precious opal which is almost priceless. If $CaCO_3$ does not work, try using oils, as both plants and animal decay releases oils, which would become trapped in the clay, possibly glycerine.

7.20 The Natural Drying Method
This is the money making end of opal production, if you can't

get this right, then you are wasting your time.

Precious Opal without the defects of cracking or crazing is very highly sought after and will fetch a very high price per carat if the colours are good.

Len had only part of the idea right, forming opal in clay. Nature however uses clay as a drying mold, squeezing out excess water as it dries, and closing up the capillaries that are formed by the escaping fluids. When the clay surrounding the opal is dry, the opal itself should be dry. The employment of lipids as elastomers, i.e. as anti cracking agents for the drying clay could form part of this method.

This hint in itself is worth millions of dollars. Dry your opal in clay.

If you choose to dry opal in air it could be possible, but would need to be in a pressure chamber. Air pressure forced into the chamber would act very much like a mold. This would be experimental, keep a written record of pressures used as trial & error would need to be used until you achieve the best result. Water pressure is already being used, in what are known as hydrostatic chambers.

If drying in clay you will have to cut and polish using this method but you will achieve a superior opal, which will be highly marketable.

7.21 Analysis of Speleogenesis

Already described in chapter 3 it has already been shown that to dissolve clay that carbonates are needed to react with the sulfuric acid producing Carbonic acid which rapidly dissolves the clay into two gels; Silica gel; and Alumina Hydroxide [$Al(OH)_3$]. The author has proved this through experimentation. The reaction is exothermic, so should not be performed in cold glass bottles. This reaction may well be compatible with the Fenton reaction as the carbonic acid is short lived degrading rapidly into Hydrogen and CO_2 but will dissolve the clay before it degrades.

Carbonates tend to inhibit the Fenton reaction but once the carbonic acid is degraded then the carbonates are no longer in solution.

Possible ingredients; water, Clay [source of Silica & Alumina], NH_4CL, FeS_2 [Source Fe_2^+ & Fe_3^+ & H_2SO_4] Dolomite [Mg & Ca Carbonate sources] & Potassium Feldspar. It is highly likely that this reaction could produce opal in nature, with the help of microbes. The Acid Sulfates & neutral chlorides are common elements in natural opal genesis.

7.22 A Caution
The opal dirt formula given by Elizabeth Smith is an untried and unproven formula, as far as the synthesis of opal is concerned. It is probably best to use the real thing, opal dirt from the opal fields.

If you want to produce a certain type of opal, travel to the opal field that produces the type of opal you want to replicate and get your opal clay there. If you want black opal, you need to go to Lightning Ridge in NSW and obtain your opal dirt from a black opal producing field.

Coober Pedy produces mostly white opal. If you have plenty of time to travel it would be good to gather a variety of opal dirt with different characteristics. Opal miners jealously guard their claims, so always

ask permission before wandering onto an opal dig, don't take unnecessary risks.

Sedimentary Opal Genesis
Author Proposes New Theory

The Opalization process is driven by natures organic decay cycle providing essential ingredients as the organic chemicals needed become available as the decay breaks down organic matter to its simplest form which is the individual chemicals such as, Ammonia, Acetic Acid, Water, Ethanol and other organic chemicals including fats oils and waxes, and humates.

Amino Acids are a large proportion of the organic matter. Amino Acids provide both Amines and Carboxylic acids. Amines decay to ammonia, and the Carboxylic acids decay to Acetic Acid which itself will decompose to Carbonic Acid, Carbon Dioxide and water. Some Amino acids are known to initiate the growth of larger silica sphere sizes, these Aminos contain more than one amine group, such as:

- Lysine
- Histidine
- Glutamine
- Tryptophane
- Asparagine
- Arginine

Some Amino Acids play specialist roles, such as Taurine [$C_2H_7NO_3S$] which is a Sulfonic acid. It is known that Sulfonic Acid creates uniform sphere sizes.

Amino acids are broken down by Enzymes, Ammonia, Iron compounds, sulfuric acid and by bacteria which accelerate the decay process.

The essential role of bacteria is to speed up the process of organic decay plus the fact that they cling to the surface of opal spheres, on which they dwell and excrete ammonia and ethanol directly to the surface of the individual silica spheres helping the growth of these silica spheres. Chemicals in solution would not be this effective. This bacterial role continues for the life of the bacteria. Many bacteria actually get trapped inside the opal as it forms because they remain on the surface of the silica spheres.

Just as synthetic opal manufacture requires a precursor which is Ethyl Silicate, natural opal formation requires a type of precursor which is biogenic silica. This biogenic silica is found in Diatomaceous earth, high silica chalk [Diatoms] and in all organic things to strengthen hard structures. Plants use it to strengthen the fibres, and animals use it in their bones or shells. I am not suggesting that the entire opal mass is made of biogenic silica, but I am saying that it is the initiator or precursor and that it has the ability to draw in silica from the soils in which it grows.

Sedimentary Opal is the resulting byproduct of natures organic recycling progam. The vast majority of the process is one of dehydration which involves ion exchange and polymerization. The opal itself is hydrated probably by the actions of carbonic acid.

There appears to been two main types of sedimentary opal growth methods in Australia. These methods operate differently but still retain some common features. In order to synthesis opal it is only a matter of copying one of these methods from nature. Let's define the differences between the two methods.

The two main Natural methods are Hydrothermal and

Diagenesis.

Hydrothermal

The hydrothermal process simply means heated water. This process is fully covered in the my Report "Opal Synthesis, Step-By-Step".

This is the process by which much of the opal in Australia formed. It has been referred to as the Mound Spring method and is described in some detail by Byron Deveson, of Canberra, Australia.

In the laboratory it is referred to as the colloidal method, or more specifically as the Sol-Gel method, due to the fact that it is a type of polymerization process. This method uses well decayed organic matter, fine particles broken down into their ultimate chemical form as source chemicals for the opal genesis process.

The Len Cram process is a prime example of man copying the hydrothermal process. Len places his ingredients into a clay mix, much the same as happens when silica spheres precipitate into the soft clay at the bottom of the mound springs. Len's secret ingredients were Saline Mound Spring water, and Acid Sulfate waters [Acid Mine Drainage] which naturally contain bacteria. Acid sulfate waters tend to be high in dissolved silica.

It is clear by Len Crams experiments that the TEOS was only an initiator and that the bulk of the silica from the clay was drawn into the Opalization process. The small amount of Silica that went into the Silcrete, was rejected by the process probably because it was a different crystal phase, making it either insoluble or too large in particle size to become opal.

Diagenesis

Diagenesis is a scientific term which means burial and pressure. The pressure referring to the weight of the burial matter pressing down on the object. There is another name for this method, it is called polymorphism. In this process, bulk organic matter forms opal within the hard structures of the animal or plant.

The organic matter provides all the fluids necessary for the process.

Examples of this process are "[1]The Opal Miner's Cat"and also fence posts of the Australian farmer that were in part converted to precious opal.

There is reason to believe that precious black opal results more often from diagenesis than other methods, which suggests an organic link to the black body colour, the background colour of black opal. It is my opinion that this colour is related to humic substances, such as Humic or Fulvic acids which are themselves a rich black colour, or possibly to Sulfonic acid which is an organic acid.

Polymorphic Opal Formation

In the mound spring process where opal forms in the clay sediment at the bottom of the springs, there is no control of the final shape that the opal will form. However, in polymorphism the opal forms an exact replica of the original structure. This is clearly seen in the way that opal fossilizes bone, shell and wood to their exact original shapes even to very fine details. Diagenesis produces polymorphic opal, or Pseudomorphs as they are often called. This type of opal could be called molded opal as it conforms to the original shape of the bone, wood or shell. Whilst hydrothermal opal is truly amorphous, without specific form, but choosing its own shape.

A New Perspective on Polymorphism

This occurs when objects such as Wood, Shells and bone are opalized. Len Cram and others have suggested the current view that opal formation is one of ion exchange. The current assumption is that the silica must come from an outside source and therefore must involve a mechanism such as replacement, or ion exchange. This would account for silica being in the place where shell bone or wood should be.

However, I present another possible mechanism that may not require an outside silica source. You see what has been overlooked is the Biogenic silica particles already present in these materials. As dissolution occurs, the silica particles grow and fill up the space once filled by cellulose, Calcium carbonate or Calcium phosphate. It is ion exchange but it is an accelerated by the growth of silica spheres. As these original materials dissolve into solution, and are drawn into the clay by ionic attraction, the silica particles expand to fill their place. This must happen very quickly for the silica to maintain the original detail of the wood, shell or bone.

It is not only ion exchange taking place; it is dissolution of calcium and dissolution and expansion of the silica. This process probably happens so quickly that some water and contaminants get trapped in the silica helping to form opal. This sudden growth of the opal spheres forces calcium and other ions from the original material accelerating the ion exchange process. Growth assisted ion exchange.

It must be remembered that the silica will expand in size, with some particles expanding over 100x their original size. This is often referred to as Ostwald Ripening, or aging. Now it becomes easy to imaging how the silica can expand to engulf the whole piece of wood, shell or bone. It literally grows in place. Although some dissolved silica may come from the outside, sourced from the surrounding clay and incorporated into the opal structure by ion exchange.

Figure 21: Silica Sphere Growth

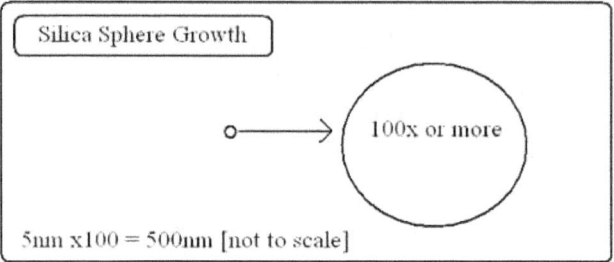

5nm x100 = 500nm [not to scale]

This can all be seen in the decay model with Ethanol and Acetic acid being the prime solvent. Ammonia would add to the power of this solvent and dissolve alumina from the clay which is attracted to the opal gel, and acts as a drying agent.

The reactants are already in place, the bone contains the chitin which is dissolved to release glucosamine, then glutamine, which forms the ammonia. Acetic acid dissolves all the fats, oils and waxes giving the mixture a higher viscosity which helps in the sphere growth process. Like ions are attracted to like ions, that is Calcium ions are attracted to calcium ions forming deposits.

In this new scenario the clay itself serves two purposes which are to act as a liquid proof barrier giving opal fluids time to produce the opal gel. It is also the source of Alumina Hydroxide which performs the function of a desiccating agent, which dehydrates to Aluminium oxide an even better drying agent for silica. The clay then dehydrates causing excess fluids to be drawn out first. As dehydration continues the clay begins to dry shrinking around the opal gel literally forming a pressure mold capable of squeezing excess water from the opal and closing up cracks even the micro cracks. It remains uncertain if the clay adds silica to the opal process due to possible differences in sphere sizes.

Biogenic silica then is the natural precursor to opal genesis, however we know that silica from the surrounding clays is often required to complete the Opalization process. This may be true of polymorphism also, however it will depend upon the amount of biogenic silica already present in target material.

Bone, wood and shell all contain silica as a strengthening agent but is there enough of this biogenic silica to fill the whole volume of the opalized piece. We can only give a diagrammatical description of how the growth process would affect the target material. Is there enough biogenic silica to cause the opalization by itself or if silica from outside is required to finish the task. The diagram or "Figure: Silica Sphere Growth" gives an indication of how this would affect the volume of the target, bone, wood or shell.

If the bone contains approx 1% silica by volume then it can be argued that if the silica growth is approx 100 xs by volume, it becomes clear that biogenic silica alone is enough to get the job done. Otherwise outside silica may be required.

1% x 100x = 100%

Vol x growth = total volume.

This growth in Silica volume would result in a pushing force that helps the ion exchange process, clearing out the other elements of the bone, calcium and phosphates that this Ion exchange is clearly responsible, in allowing the silica to penetrate the entire volume, whether it be a piece of wood, a shell, bone or a tooth. The role of Ion exchange is not diminished by the growth of the silica volume.

Conventional views on Polymorphism

This may be considered as a different type of sedimentation, where the particles do not simply settle on the bottom of the tank but are attracted to bone wood or shell which they penetrate by a process of intensive ion exchange [replacement] before or concurrently with polymerization, resulting in an almost exact replica of the target object.

Comparison of Natural Methods

Figure 22: Comparison of Natural Opal Genesis Methods

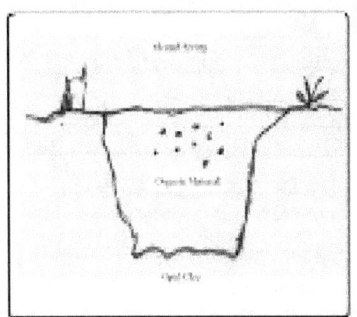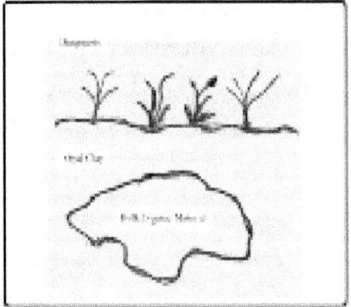

The Hydrothermal Method

The Len Cram Method is a classis example of the Hydrothermal Method which is also exemplified by the Mound Spring Theory of Byron Deveson.

Examples Include:

- Mound Spring Opal, Most South Australian Opal including the Coober Pede area.
- Len Cram Opal
- Hot springs

- Geothermal springs

Microbes played a pivitol role in both of these methods. In The Mound Spring Method Silica spheres formed in the water, and grew to opal sphere size in the water. Microbes in the water fed on microscopic decayed organic matter [Dissolved Organic Matter, DOM].

The Diagenesis Method

In simple terms Diagenesis could be called, The Burial Method

Polymorphism is typical of this method. Much of the opal of

Lightning Ridge has been formed by the polymorphism.

Examples include:

- Lightning Ridge Opal
- Virgin Valley Opal
- The Miners Cat
- The Farmers Fence Posts

Virgin Valley opal is included here, although the starting materials were of volcanic origin, the opal did not form until later when the organic materials that were buried by the volcanic ash began to decay. It is a sedimentary opal deposit.

Microbes in this model feasted on bulk deposits of organic matter buried underground, in opal dirt.

The decaying organic matter provided all the organic chemicals needed plus the water also.

A Third process "Spieleogenesis"

There is another natural opal forming process which is also a cave forming process. It is thought to be a minor opal producing process called "Spieleogenesis", where sulfuric acid comes in contact with carbonates, and dissolves the carbonates, producing carbonic acid which can then dissolve large amounts of silica. It is then very much like the other processes.

Similarities

All methods require organic chemicals both acids and bases, and all require, bacterial activity [catalyst for decay, direct application of chemicals to silica spheres], a form of easy soluble silica ie biogenic silica which can only be sources from organic matter, and maybe the only form of natural silica with small enough particles to initiate the process. Biogenic silica is found in both plants and animals.

See the following Diagram and explanation below.

Summary Basic Opal Formation Steps

Natural Sedimentary Process

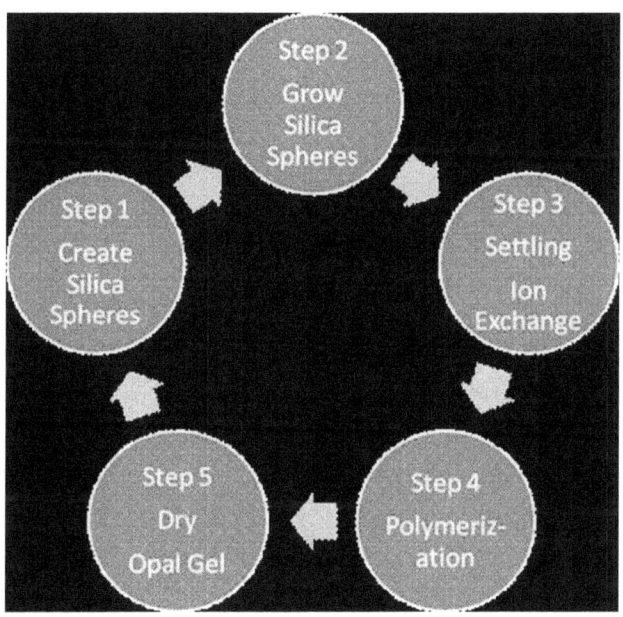

Bacteria are present due to organic decay

Note: This Diagram is simplified and therefore not as detailed as the previously shown 8 step diagram.

Overview of Process Diagram

This process is generic to both Natural Sedimentary Opal formation processes and also to man made or synthetic opal generation.

Step 1

The Opal formation process begins with the formation of Silica Spheres which can happen as a natural result of the dissolution of Biogenic silica or by the hydrolysis of Ethyl Silicate.

Step 2

Silica Spheres begin at approx 5nm and must grow to between 100nm and 700nm in size, as this is the approx. size range of opal spheres. The growth agents for this process is amines from amino acids which decay to ammonia. Bacteria apply this ammonia directly to the surface of the spheres along with Ethanol.

Step 3

In Nature Settling happens within the opal clay, this applies to all the sedimentary processes. As the Silica spheres begin to settle in the clay, the self alignment begins based on silica spheres having a positive charge at one end and a negative charge at the other end. The Alignment is negative to positive like a magnet. During the settling process ion exchange takes place, more soluble silica can be brought into the process by the ion exchange process and other materials can be rejected.

Step 4

The polymerization process also takes place during the settling phase. The self aligned spheres first form strands, when enough strands form they begin to bundle together, and cross-linking occurs.

Step 5

The whole process is one of dehydration and as the formation dries out, the opal gel contained within the clay also dries out. This may happen over a long period of time.

In opal synthesis special drying techniques are used, some more effective than others.

Summary of Natural Opal Formation Methods

Some possible ways in which opal may form in nature are:

- Volcanic [Not relevant to this topic].
- Mixing of Acid Sulfates and neutral chloride solutions. [Examples: Mound Springs, or Hydrothermal]
- Organic matter buried in opal dirt [Diagenesis,Example: Polymorphic Opal].
- Speleogenesis [Opal formed in caves is often by this method].

Shared Credit

Much of the credit for research into Mound Spring opal must go to Byron Deveson who did such an excellent job not only of the field research but the excellent way in which he conveyed this information in his paper *"A new Opal Model"*.

References:

[1]"The World of Opal" by Allen W. Eckart.

[2] *"Black Opal Fossils of Lightning Ridge"* by Elizabeth Smith

[3] *"A new Opal Model"*
Mound Springs as Opal Source
Byron Deveson
Canberra 2006

The End

ABOUT THE AUTHOR

Hi! I'm Richard Shepherd. Since my high school days I have been fascinated with Opal, it's great beauty, the variety of colours and the mystery that surrounded its formation.

My research into opal and its formation began around 1972 when our science class at school studied opal but failed in many respects to explain the origins of opal and its formation. Information on the subject was old and dated and made little sense when truly analyzed.

I had studied polymerization in science earlier that year and when handed a piece of opal to observe, it was clear to me that this opal was in fact a polymer.

This led to my decision to collect information on opal and to do my own research and to discover the real origins of opal for myself, to answer the many questions that I had been asking about opal myself.

Having discovered early in life that I possessed certain talents which would help me in this goal. These talents included an ability to do research, to analyze information, a gift for problem solving and invention, also an ability to create concepts from design to story concepts. My imagination just seemed to exceed that of anyone around me.

This book is a testimony to my talents, a type of resume if you wish but also a journey of discovery. It is the culmination of a dream to write a book, not just any book but a technical book written in text book format, making it easy to search and find the information that you seek.

Now, at the age of 53, I have found myself back in the dating game. My deaf wife of 22 years suddenly decided that we were not communicating well enough to remain a married couple. I have been separated from her since January 2011. Our differences are irreconcilable.

I am now looking for another life partner. The woman that I am looking for should be between the ages of 32 – 45 years old. In particular you should be a born again Pentecostal Christian preferably with a Christian heritage. A woman who loves life, has a positive outlook, and is free to enjoy life. You should still enjoy a glass of wine or a little alcohol. You should enjoy cooking and be hospitable.

I have 4 daughters ranging in age from 11, 14, 19 & 21 year old. At my age not looking to have any more children. If you cannot have children that would be an advantage. If you have a healthy libido, that would also be an advantage. If I could make a lonely woman happy that would be wonderful.

You should be a down-to-earth woman, easy to talk to, intelligent, mild mannered, forgiving and patient, with lean figure and attractive with a good sense of humour would be preferred. You should dress well and look feminine, not like a Tom-boy. Important that you understand and speak English well. You should be very healthy, have good hearing and good sight. Must want a husband and be affectionate. Interested? Email:
RichardShepherd678@yahoo.com.au
In subject line type "Flirt"

My interests include, Swimming, playing Tennis, Table Tennis, Walking, Dancing (Ballroom), B.B.Q's, Dining out, Research, Technology, Photography, Design, Innovation and Invention. My plans for 2014 include a change of career to invention, as I have many of my own designs to develop.

Photo's will be available on request to interested women.

Good-Bye for now.

BUSINESS PAGE

DISCLAIMER

Do not try to make opal at home. Your chances of success are almost zero without knowing the quantities required and having a quick drying method which are not included in this book.

Ethyl Silicate is Caustic and can be harmful if not used with the proper safety precautions. Anyone who wants to use it should get MSDS information about it before even thinking about using it.

The author will not be held responsible for any damage or injury resulting from the use or misuse of any of the formulas, or methods in this book. Information in this book is provided for the purpose of Information only.

THE BIRTH OF OPAL

www.ingramcontent.com/pod-product-compliance
Lightning Source LLC
Chambersburg PA
CBHW051436170526
45166CB00001B/14